Darwin's Legacy

Darwin's Legacy

What Evolution Means Today

JOHN DUPRÉ

University of Exeter

OXFORD
UNIVERSITY PRESS

OXFORD
UNIVERSITY PRESS

Great Clarendon Street, Oxford OX2 6DP

Oxford University Press is a department of the University of Oxford.
It furthers the University's objective of excellence in research, scholarship,
and education by publishing worldwide in

Oxford New York

Auckland Bangkok Buenos Aires Cape Town Chennai
Dar es Salaam Delhi Hong Kong Istanbul Karachi Kolkata
Kuala Lumpur Madrid Melbourne Mexico City Mumbai Nairobi
São Paulo Shanghai Taipei Tokyo Toronto

Oxford is a registered trade mark of Oxford University Press
in the UK and in certain other countries

Published in the United States by
Oxford University Press Inc., New York

A catalogue record for this title
is available from the British Library

British Library Cataloguing Publication Data

Data available

Library of Congress Cataloging in Publication Data

Data available

Typeset in New Baskerville by
Footnote Graphics Limited, Warminster, Wilts
Printed in Great Britain
on acid-free paper by
T. J. International, Padstow
and bound by
Biddles Ltd., King's Lynn, Norfolk

ISBN 0–19–280337–9   978–0–19–280337–5
ISBN 0–19–928421–0 (Pbk.)   978–0–19–928421–4 (Pbk.)

1 3 5 7 9 10 8 6 4 2

For Regenia

Preface

A couple of years ago I received an invitation from Shelley Cox, then of Oxford University Press, to contribute to a series of short books addressing big issues in philosophy. I was working at the time on issues connected to evolution, and it occurred to me that an essay on the broad implications of evolution would very well fit this bill. The series did not, in the end, come to pass, but the present book was a consequence of its conception. That big issues—the existence of God, our understanding of human nature, and our relation to other creatures—are affected by ideas on evolution is hardly disputable. The attempt to express my views on these matters succinctly and in a way accessible to a wide audience has been challenging and rewarding.

The book has undoubtedly benefited from discussions with and writings of many scholars, colleagues, and friends over the twenty-five years or so that I have been concerned with issues in evolutionary theory. I shall not try to list them here. More immediate debts are to audiences in Ghent, Bristol, and Exeter who responded to a paper presenting some ideas developed in Chapters 3 and 4; to Anya Plutynski who sent me some useful comments on a written draft of that material; to Jonathan Kaplan who saved me from some errors in Chapter 7; to Shelley Cox and Emma Simmons, at Oxford University Press, for gentle but firm encouragement to get on with the project; and to Marsha Filion, Shelley's successor at the Press, for detailed and valuable advice on making the book, especially the opening chapters, read better.

I am grateful to many colleagues at the University of Exeter for contributing to the congenial and intellectually stimulating environment in which the book was written. It was completed as I began my tenure as Director of the newly established Economic and Social Research Council Centre for Genomics in Society, and I am very grateful to the ESRC for their support. As always my greatest debt is to my partner, Regenia Gagnier, for continual intellectual stimulation and dialogue and much else besides. The book is dedicated to her.

Contents

1

Introduction

My occupation is still an unusual one in the UK, though commoner in the US, where I have spent most of my professional life. My training is in philosophy, but what I study is biology. Like every other philosopher of biology of whom I am aware, I have spent much time thinking about evolution. Philosophers of biology typically spend a good deal of their time looking at rather small questions about evolution: Does natural selection act on genes or individual organisms? Does evolution proceed at a steady pace or in intermittent bursts? To what extent are organisms optimally fitted to their environments? But surely there are big questions too. What does evolution tell us about ourselves? About our place in the universe? About God? In this book I attempt to distil the views on these questions that I have formed over several decades into a slim volume. Put most generally, the question I want to address is, what does evolution tell us about ourselves and our world? Or put another way, why should we (non-biologists) care about evolution? The answer I shall offer in this book is that evolution does indeed have momentous consequences for our view of ourselves and our place in the universe, but that it does not have the kind of consequences most widely advertised today. In particular, it is of limited use in illuminating human nature.

Few people doubt the importance of the theory of evolution. Its development is, at the least, one of the paradigmatic achievements of Western science. One can study the theory of evolution as an example of how it is possible to gain profound insight into our world. But of course this knowledge is not merely interesting as an example of knowledge; it has also profoundly affected our understanding of ourselves and of our place in the universe. While seen by some as providing a novel account of God's ways of world-making, others have seen the theory as the last essential element in a naturalistic and materialistic view of the universe, and as thereby removing the last hiding place for a God or gods. At a more mundane level, explaining how aspects of human nature grew out of the exigencies of our evolutionary history has become one of the most reliable ways of writing a best-selling book. Others have accused the writers of these best-sellers of pseudo-scientific story-telling or worse. These debates have been some of the most heated in recent intellectual life. This hardly needs to be said for the exchanges between creationists and evolutionists, but the arguments internal to biology in which well-known public figures such as E. O. Wilson, Steven Pinker, or Richard Dawkins have lined up against biologists such as Richard Lewontin, Steven Rose, and Stephen Jay Gould, have been hardly less vitriolic. This book will provide an examination of these debates and offer opinions about which side of them is most credible.

Belief and scepticism

There has been tension between evolution and theology ever since Bishop Samuel Wilberforce famously, if perhaps apocryphally, enquired of Darwin's great champion T. H. Huxley on which side of his family he claimed descent from an ape. Today fundamentalist Christians still manage to make themselves look ridiculous by attempting to remove the teaching of evolution from school curricula. Most Christians are much smarter than this, of course, and increasingly claim that there is no great difficulty in reconciling evolutionary ideas and Christian belief. Such a view has been defended by prominent biologists and philosophers. I reject this accommodation. In the unfamiliar company of Richard Dawkins, the most prominent contemporary scientific atheist, and fundamentalist Christians, I believe there is a clash, and that the religiously minded are right to be fearful of the general acceptance of evolutionary thought.

Scepticism has always been one of the greatest intellectual contributions provided by philosophy. Many philosophers have admired science not so much because of its discoveries, but rather because of the cautious, provisional, and even sceptical methods employed by the best scientists. An extreme example of this attitude was that of perhaps the most famous philosopher of science, Sir Karl Popper. Popper didn't think that any scientific claims should necessarily be held as true. What he admired was the disposition of scientists to attempt to refute hypotheses. Popper no doubt exaggerated the importance of this disposition in elevating the process of attempted refutation to the basis of all genuinely scientific methodology, but the attitude expressed by his thesis is an admirable one.

I, on the other hand, do maintain that science can sometimes accumulate enough evidence for its claims to make them almost impossible to reject, and I believe that certain broad evolutionary theses have earned this level of credibility. (Popper, at one time, claimed that evolution was unfalsifiable and therefore not a genuine scientific theory, though he later retracted this assessment.) The crucial idea here is *evidence*. Another philosophical tradition to which I subscribe is empiricism, the commitment to seeing knowledge as ultimately based on evidence from experience. The basic idea that any limits to our scepticism, any propositions about our world that we are entitled to believe, must be based on evidence seems to me absolutely correct. And, as I argue in detail in Chapter 4, this principle rules out any well-grounded theology. Simply put, we have evidence to support evolution, but none to support belief in a deity. This, I suppose, is one good reason for caring about evolution.

Evolution and human nature

My scepticism about religious claims is not based on a dogmatic commitment to the claims of science. On the contrary, and this is perhaps where a philosopher's view on the subject is liable to be rather different from a biologist's, I take a sceptical attitude to many supposedly scientific claims too. In particular, I consider contemporary attempts to treat evolution as the key to all mythologies, the route to deep insight into human nature, to be profoundly mistaken. Such ideas are currently very fashionable, most prominently perhaps in the work of so-called evolutionary psychologists

In Chapter 5 I consider the relation of humans to the rest of the animal kingdom. Much religious thought tries to erect an impassable barrier between humans and other animals: we, but not they, have souls. At the other extreme, evolutionary psychologists often like to insist that we are really just one species of animal like any other, and exploit this claim in defence of various theses about human nature. So, for instance, the argument that women are naturally inclined to seek out males with substantial resources might be introduced with a discussion of the grey shrike. The male birds of this species assemble collections of food and household knick-knacks (feathers, pieces of cloth) to attract acquisitive females. From this example it is insinuated that men willing to provide a nice house in the suburbs with pleasant curtains and a well-stocked larder will be more attractive to the human female. Or, a more sinister example, descriptions of drakes lurking behind bushes and leaping out to sexually assault passing ducks is offered as evidence that men, too, may have a natural disposition to rape.

Here I find myself unexpectedly closer to the side of the angels. Whether or not we all believe in immortal souls, the inference from animal to human behaviour is a suspect one. It is quite generally true that such parallels are little more than ornamental. The fact that a feature evolves in one species simply shows that it can evolve, and the fact that some other species lacks that feature shows that it may fail to evolve. The details of the behaviour of unrelated species are therefore of little relevance to understanding a particular species such as our own.

But it is also important that we do not short change what is extraordinary about our own species. Our common lack of immortal souls does not prevent us from being very different in important ways from other creatures we have encountered. Our cultures are orders of magnitude more complex than any non-human social systems of which we are aware, and surely a key to the possibility of such cultures is the unparalleled complexity of human languages. These are hardly original points, but they are often and deliberately obscured by much popular evolutionary theorizing. Our words, in fact, may be our best chance of immortality.

Chapter 6 addresses evolutionary psychology head on. Evolutionary psychologists generally see their main opponents as social scientists, liable to adhere to something called the Standard Social Science Model (SSSM), according to which the human mind is a product of culture wholly unconstrained by any kind of human biology. As Steven Pinker puts it in a recent book, they believe that the human mind is a 'blank slate'. In opposition to such a picture, evolutionary psychologists argue that human nature is far less variable than is generally supposed, and that it in fact consists of a large number of mental modules, designed by natural selection to generate evolutionarily optimal behaviour in response to environmental cues. It is a project that appears to resonate strongly with the contemporary Zeitgeist. But it is also, in my opinion, deeply misconceived.

It is notoriously dangerous to suppose that understanding how something came about is the right way to understand what it does or how it works. It would be wrong to infer from the fact that George W. Bush was descended from powerful politicians that he was well suited to be a politician, and equally wrong to infer from the fact that he spent a large part of his life in the oil industry that his policies were directed mainly to furthering the interests of the oil industry. Both of these claims need to be investigated on their merits. They may suggest causal processes that might have led to the suggested characteristics, but a good deal more is needed to establish that these characteristics have been realized. None of this is to deny that history matters. As we all know, those who are ignorant of history are doomed to repeat it. But history does not only provide insight into human nature but is also in part a determinant of human nature. It has become increasingly common, for example, to think of the demand for democratic modes of government as a fundamental expression of human nature. But if it is, it only became so through a long history of argument and struggle.

My crucial point here is that given that history matters, then we must ask which part of it matters. Evolutionary
 biologists claim that the history that matters, the history that laid down the fundamentals of human nature, is the grand sweep of evolutionary time. Something as historically ephemeral as a preference for democracy, they will insist, is

too superficial to count as part of human nature. My own view is that more recent history is generally a lot more relevant. Certainly a great deal of evolution of the brain happened long before there were any beings at all like humans, and without this evolution there could have been no human mind. But to understand the human mind, or human nature, we need to look at much more detailed ways in which particular cultures have developed and coevolved with the people who live in them.

Bringing together my last point with the earlier discussion of evolution and theology, I suggest that two extreme timescales encompass the parts of history that really matter to us. One is the very long term. It was enormously important, first, to discover that there was a very long term rather than the few thousand years discerned by ancient biblical exegetes. It is even more important that we now have a general picture of the kinds of things that did, and did not, happen in the aeons of time of which we are now aware. But, finally, it is equally important to appreciate the variety of the kinds of human behaviour that developed in the last few thousand years.

It is no doubt of great interest to try to understand what the creatures that evolved into modern humans were like. This project leads evolutionary psychology to emphasize the Stone Age, the last million or so years of evolution. This is certainly an important part of pre-human history, but there is a good deal less to learn about modern humans from studying this period than evolutionary psychologists suppose. A crucial part of answering the question why we should care about evolution, it turns out, is deciding which bits of evolution we should care most about.

A final issue I address is the relevance of evolution to contemporary classifications of humans, and in particular the classifications of humans that have caused the greatest controversy and the greatest suffering: race and sex. Apart from occasional disturbing mutterings about the evolutionary advantages of xenophobia, evolutionary psychology has not had much to say about race. And there is much to be said for leaving things that way. Nonetheless we have a pretty good idea what the evolutionary history of racial differences probably amounted to, and it will surely do more good than harm to be aware of this when discussing race. Sex is quite another matter, and has indeed been the greatest focus of research by evolutionary psychologists. I have mentioned women's supposedly economic approach to sexual partnership and men's alleged disposition to rape. In this area there is no discernible limit to the bizarre claims that are sometimes made on evolutionary grounds. Ben Greenstein, a biologist who has studied hormones, writes on the dust jacket of his book *The Fragile Male*, 'First and foremost, man is a fertilizer of women. His need to inject his genes into a female is so strong that it dominates his life from puberty to death. This need is even stronger than the need to kill.' This is an extreme example, but it dramatizes the problem. Evolutionary thinking is here misused to paint a crude and often nasty picture of human nature and one that is receiving wide circulation. No such vision of human nature follows from the facts of evolution. Later in the book I shall explain exactly how the arguments alleged to lead to such a vision go wrong.

The last point brings me back to the earlier chapters of the book that I haven't yet described. In Chapters 2 and 3 I try to explain what the theory of evolution is and what it does or, more specifically, what sorts of things it can explain. This is the part of the book that will be most likely to annoy my professional colleagues, as it sets out sometimes controversial views on the interpretation of the scientific theory, and it is likely to be the heaviest going for the general reader, as the subject matter is at some distance from the direct implications of evolutionary thinking that are the main topic of the book. Here I beg the reader for patience. Obviously enough, we cannot decide what evolution entails without a fairly sharp conception of what it is. It turns out that here, too, some rather simplistic ideas have gained considerable popular currency. Our understanding of evolution is, however, continuing to develop, and to develop in directions that enable us to see very clearly the difficulties encountered by the kind of speculations that will be the subject of later chapters.

This, then, is a philosopher's view of evolution. It is a philosophical view grounded in scepticism and empiricism. By scepticism I do not mean that I shall speculate as first-year philosophy students are often required to do about whether we might all be brains suspended in nutrient fluids with experiences fed into our cortices by hyper-intelligent aliens. Rather, this book will always question whether we have good grounds for believing those things that science, authority, tradition, and so on, encourage us to believe. Empiricism provides the standard to which beliefs should answer. If we are capable of finding out what kind of world we live in, surely the best way of doing so is through our experience of it. Science has always aspired to empiricism but has not always lived up to that aspiration.

This is an austere conception of how we should decide what to believe, and will perhaps not appeal to many. But the exciting lesson of science is that it is possible to learn about our world and to do so at a remarkably deep level. It is vital that this success should not tempt us to take short cuts with enquiry that will lead us back into the dogmatism of pre-scientific world-views. It is in the hope of contributing to the avoidance of this danger that I advocate this sceptical approach to one of our greatest scientific achievements.

2

What is the Theory of Evolution?

Overview

The 'theory' of evolution is most often used not so much to refer to a scientific theory at all, but rather to a set of very general propositions. The core proposition of evolutionary thought is just the simple fact that life on Earth evolved. Complex forms derived from simpler forms and all, or certainly most, life forms share common ancestors. The fundamental idea underlying these claims is descent with modification. These statements are beyond any serious question. Some general outlines of the more specific relationships between organisms, for example that all mammals share a common ancestral species, are also known with great confidence, though perhaps are too local to belong among the core propositions. Some would include within the core the proposition that the earliest life evolved from non-living materials. However, following Darwin, I do not take this to be a necessarily integral part of the theory of evolution. If, as some respected scientists have argued, primitive life forms arrived from outer space, this would not alter much of what we believe about subsequent evolution. I shall return to this question again at the end of the book.

Needless to say, this simple set of propositions is not the end of the story, and matters soon become more problematic. Many of the problems might be avoided if we could somehow outlaw the phrase 'the theory of evolution'. A relatively trivial reason for this is that the word 'theory' is often used in popular parlance to import a high degree of speculativeness, as in 'that's just a theory'. This connotation is often exploited by Christian fundamentalists, arguing that their theory (biblical creationism) has as much claim to our attention and the attention of our children as ours (evolution). There are certainly speculative elements concerning evolution, but there are other elements, notably the core propositions and some more peripheral facts about the relations between particular kinds of organisms, that are as unquestionably true as anything that science has established. A main aim of this chapter will be to distinguish those parts of the theory about which there remains no serious question from those around which there remains lively scientific debate.

A second and more serious problem with speaking of 'the theory of evolution' is with the definite article. Evolution encompasses a complex set of beliefs with varying degrees of certainty. 'The' theory of evolution suggests a unified whole and perhaps even one that must be swallowed or rejected in its entirety. And in more technical areas of philosophy of science, the reference to evolution as a theory has suggested a parallel with such things as the theory of relativity, or quantum theory. These theories in physics have been more traditional objects of philosophical enquiry, and it has taken decades of work to realize that the articulation of an account of evolution was a very different kind of intellectual endeavour. It remains controversial whether such physical theories can, as an earlier generation of philosophers of science argued, be usefully thought of as axiomatic systems, but it is quite clear that evolutionary theory cannot.

Evidence for evolution

I have said that descent with modification and the relatedness of different biological kinds are facts. I do not, of course, mean to assert dogmatically that they could not turn out to be false. Perhaps, as philosophers occasionally like to speculate, we are all nothing but brains in vats of nutrient fluid, the playthings of malicious aliens. But barring the kinds of sceptical hypotheses that threaten all our claims to knowledge, these are facts as well established as any. They are established by the overwhelming convergence of evidence. First, there is physiological evidence of related structures. One classic example is the structure of the mammalian forelimb. The wing of the bat, the flipper of the whale, and the human arm all share exactly the same arrangement of bones, but organized for quite different functions. At the microscopic level, all organisms share the same relations between DNA sequence and the structure of the amino acids, the assembly of which the DNA directs.

The overwhelmingly compelling explanation of this and countless parallel examples is descent from common ancestors. The same mammalian bone structure has diverged to serve the quite different forms of life of different organisms.

Second, there is the evidence of fossils. Rock strata, which can themselves be dated by a variety of physical techniques, disclose sequences of organisms that change from currently unknown forms in ancient rocks towards forms in more recent rocks that are increasingly similar to creatures alive today. The pattern of descent revealed by fossils is wholly consistent with the pattern of relationship suggested by physiological comparison.

And third, there is the evidence of biogeography, the geographical relationships between organisms of different kinds. The classic example here is of Darwin's finches. These small birds have occasionally been blown from the South American mainland to the Galapagos Islands and thence to other islands. In the absence of indigenous competitors these birds have rapidly evolved to an extraordinary range of different ways of life, normally the province of specialized groups in more competitive continental environments. So there are closely related species of wood-pecking finches, insect-eating finches, nut-cracking finches, and so on. It would be something of a mystery why God would not have provided these islands with proper woodpeckers; evolutionary thinking has a ready explanation for the oversight.

These different kinds of evidence all converge not only on the fact of evolution, but on the more or less detailed patterns of evolution. It is always possible to find alternative explanations for anything. The nineteenth-century naturalist Philip Gosse notoriously suggested that God had created the Earth complete with layers of carefully arranged fossils, perhaps as a test for our faith. But this is quite clearly an ingenious attempt to reconcile the phenomena with antecedent belief, hardly an explanation compellingly connected to empirical reality. The distinguished evolutionist Theodosius Dobzhansky famously remarked that nothing in biology makes sense apart from the theory of evolution. While I shall suggest later that in certain respects this is a considerable exaggeration, it is easy to see the point. There is so much that does make sense in the light of evolution that it is inconceivable that, in broad outline, evolution is not a fact.

The central theme of this book is that while this general fact is of enormous importance to our understanding of our place in the world, the amount of detailed understanding of biology with which this fact provides us is often overstated. At the risk of sounding pretentious, I want to say that the greatest importance of the theory of evolution is metaphysical: it tells us something very general about what our universe is like and what sorts of things there are in it. This importance depends on no more than the broad propositions just adumbrated. Now, however, I move towards the scientifically more interesting areas of evolutionary theory where controversy remains very much alive.

Natural selection

It is common for biologists to refer not just to the theory of evolution, but to the theory of evolution by natural selection. The importance of natural selection in the process of evolution was Darwin's great contribution to science. Natural selection is now widely agreed to be by far the most important factor in understanding how the modifications that occur over the course of evolution are possible. The theory is best understood through the idea of heritable variation in fitness. Fitness is, roughly speaking, the disposition to produce surviving offspring. If organisms differ in respects that give rise to this disposition—differ, that is to say, in fitness—some will leave more offspring than others. If fitness is heritable and the features that ground differences in fitness are passed on from parents to offspring, then features conveying fitness will become more common. Evolution by natural selection will therefore produce particular kinds of changes in populations of organisms, changes towards greater prevalence of features with greater fitness. To take a classic example from evolution in the nursery, if giraffes are fitter with longer necks, allowing them to reach higher leaves and survive food shortages, for example, and if long-necked giraffes have longer-necked baby giraffes, then a population of giraffes will evolve longer necks.

The crucial point is that in moving from the mere fact of evolution to evolution by natural selection, we are moving from fact to genuine theory. This is not to say that there is not a fact of natural selection. There is. There is no serious doubt that natural selection occurs, and very little serious doubt that it is of great importance in the processes of evolution. We are in the realm of theory because natural selection is the focus of an enormous amount of scientific controversy. In particular, there is controversy about exactly how important natural selection is to the evolutionary process, and there is controversy about how we should understand the process itself. Let me treat these in turn.

Darwin did more than anyone else to establish the fact of evolution because, through his emphasis on natural selection, he showed that it was possible. Many of the facts that make evolution such an irresistible explanation of the nature of biological diversity were well known before Darwin, and various thinkers advocated a version of evolutionary theory. The great difficulty was that no one had a convincing account of how the physiological changes that evolution implied could come about. Most importantly, it was unclear how living forms came to be so exquisitely adapted to their ways of life. The whale's flipper and the bird's wing seem wonderfully adapted to swimming and flying. If we want to claim that both evolved from some common ancestor that perhaps neither swam nor flew, we need a good story about how this adaptedness came about, and this is exactly what natural selection provided. Constant variation among the ancestors of birds, and selection of those that, perhaps, were least likely to injure themselves falling out of trees, might lead after millions of generations to creatures with perfect wings. Darwin's brilliant insight has subsequently been greatly refined. Most important was the integration with

Gregor Mendel's discovery of particulate inheritance.
Darwin had assumed that inheritance involved an equal
blending of the characteristics of both parent organisms.
Critics pointed out that this idea would lead rapidly to homo-
geneity throughout a population, and hence exhaust the
variation on which selection was supposed to act. Mendel's
work, first published in 1865 but ignored until simultaneously and independently taken up by three biologists in
1900, showed that this view of inheritance was mistaken.
The inheritance of the traits Mendel investigated in peas
was an all or nothing affair. The factors responsible for the
inheritance of such traits would not, therefore, be swamped
by blending with other factors, but could become established throughout a population. Subsequent research has
developed far deeper insight into the nature and source of
genetic variation and also into the processes of selection.
The result today is a richly articulated causal theory. More-
over, natural selection remains by far the most powerful—
according to many the only—theory that provides an
explanation of the adaptation of organisms to their environments.

Controversies

No serious biologist, as far as I know, doubts the enormous
importance of natural selection to the evolutionary process.
Nonetheless, there is considerable disagreement about
the extent to which natural selection is sufficient for
understanding evolution. Very heated debate concerns the

power of natural selection to produce optimally adapted outcomes, power that is not unlimited. It might be that snails would do better traversing the ground on caterpillar tracks and thereby save the expense of leaving a trail of slime behind them wherever they went. But it is exceedingly unlikely that there is any accessible evolutionary path from where snails now are to the imagined scenario. The problem can be thought of through the metaphor of an adaptive landscape. Think of a particular kind of organism as occupying a mountain top, symbolizing a position better adapted than any nearby position. Though there may well be higher peaks, to reach them evolution would have to take the population of these organisms through the intervening valley, which would require moving through less adapted positions. Since natural selection can only move the population upwards to positions of greater fitness, the path through the valley of lower fitness is impassable. These restrictions are examples of a broader class of concerns thought of as *constraints* on evolution. Though these are sometimes classified under a variety of headings (genetic, developmental, physiological, and so on), the most important are historical, a consequence of the very fundamental fact that evolution is a historical process: where one can go depends more than anything else on where one is, and on how one got there.

Doubts about the power of natural selection to overcome such constraints have encouraged some to look for other processes, perhaps even of importance comparable to that of natural selection, which would need to be recognized in a full account of evolutionary history. One interesting candidate, whose major proponent is Stuart Kauffman, is a suggested tendency to self-organization of some complex systems. Kauffman has used methods including elaborate computer simulations to suggest that systems such as complex soups of organic chemicals will tend spontaneously to produce stable states strikingly reminiscent of the chemistry of living cells. He does not suggest that this is an alternative to natural selection, as the systems that might arise from such processes will surely be filtered by natural selection to leave those that are most stable and able to reproduce themselves. But it does open up the possibility that biological order may be explained in part by processes very different from those of natural selection. Other biologists, more optimistic about the power of natural selection to overcome constraints, insist that natural selection alone is fully up to the task of explaining evolutionary history.

There are also important debates entirely internal to the theory of natural selection. Perhaps the most heated and long-running such debate concerns the question of what exactly it is that natural selection selects. This question achieved prominence with Richard Dawkins's highly visible and influential book, *The Selfish Gene.* In that book Dawkins, following seminal earlier work by G. C. Williams, argued that the target of selection was not the organism, but the gene. What natural selection accomplished was the prevalence of those genes that were most efficient at reproducing themselves. The organism, on this view, was to be seen as merely a vehicle built by genes in order to project themselves most effectively into the next generation. The theory was presented in large part as an attack on the once popular idea of group selection, the idea that certain features of organisms, particularly those that appeared to benefit conspecifics rather than the individual itself, might have arisen through selection between groups of organisms. In other words, groups in which organisms provided benefits to one another would have survived better than those composed only of selfish individuals. Dawkins argued that this process could never work, since groups with selfless, 'altruistic' individuals would inevitably be subverted by selfish mutants who would derive the benefits of others' altruism without paying the costs of being altruistic themselves. It is a curious fact, however, that this attack, whatever its merits, has no bearing on the issue of whether natural selection is better seen as acting on genes or whole organisms. The negative consequences of the excessively gene-centred view of evolution will be an important theme later in this book.

Dawkins's theory has been extremely influential, particularly on the sociobiologists and evolutionary psychologists whose work will be the subject of a later chapter. Although group selection became an unfashionable concept for a while, this was by no means the end of the debate. A number of biologists and philosophers have argued that the selection of genes is insufficient to represent fully the complexity of the evolutionary process. They have proposed that selection occurs simultaneously at many levels, including at least the gene and the individual, and very possibly also the group and even the species.

Multi-level selection is perhaps the current orthodoxy among philosophers of biology. There is, however, a more radical movement that is rapidly gaining adherents. This movement sees evolution as a continuous process and doubts whether it is possible to analyse it successfully by separating out a privileged set of objects involved in a certain stage of the process. The most influential version of this perspective is developmental systems theory (DST). Among other things, DST provides a powerful critique of Dawkins's genic selectionism or, more generally, gene-centredness. Proponents of DST insist that genes are not as special or unique as both professional and popular opinion often suggests. It is often said, for instance, that genes are the bearers of information about the organism that they help to build. More portentously, one often hears the suggestion that an organism's genes provide a plan or a blueprint for the whole organism. Such statements were often heard from proponents of the human genome project. But this sort of talk is highly misleading. As the blueprint metaphors make explicit, information often carries semantic connotations. But of course there is nothing semantic about DNA.

There is a technical meaning of 'information' in which it means only that the state of one thing (the bearer of information) provides more or less reliable predictions about the state of another (that about which it gives information). Genes that increase, even slightly, my disposition to cancer

instance. But if this is all that is meant by talk of genetic information, one could equally well speak of the information carried by a great many structural and chemical features of the cell, and even of features of the environment in which the organism develops. Sunny weather, for instance, carries information about the bright red colour to be expected in my ripening tomatoes. Genes, on this view, are simply one— no doubt a very interesting one—of the resources that the organism requires for its proper development. DST offers as the unit of selection the entire developmental cycle. Successful sequences of developmental cycles prosper and diverge. Others less well adapted to their conditions die out. Central to being well adapted is the ability to gather together and deploy the full set of resources necessary for producing the next generation or, more accurately, the next iteration of the cycle.

The most important aspect of this perspective is that it breaks down the divorce between evolution, on the one hand, and development on the other. The failure to integrate these aspects of biological thinking has often been noted. And the spuriously Janus-faced gene—at one and the same time the unit of selection and the fundamental cause of development—has served to conceal this divide. On the picture propounded by Richard Dawkins, natural selection picks out a set of genes in each generation, and the genes then determine the next set of phenotypes which are then again subjected to selection. This picture allows development to be treated as a 'black-box', the details of which are of no importance. The genes can take care of that when they have finished with the important business of being selected. If genes in fact provided blueprints or recipes for the production of organisms, this black-boxing might be acceptable. But in fact they do nothing of the kind. As just noted, the information required to build an organism is distributed over many levels of biological and external organization. Genes are only selected to the extent that they participate in complete successful developmental cycles.

Another lively debate has concerned the pace of evolution. Ever since Darwin there has been a concern about the gaps in the fossil record. Much of this record appears to consist of long periods during which quite constant forms are encountered, followed by sudden breaks, after which substantially different forms are found. It was assumed for many years that this anomalous pattern represented only the very unreliable processes by which fossils were produced. But some years ago the distinguished biologists Stephen Jay Gould and Niles Eldredge argued that this pattern actually represented quite faithfully the history of life. Evolutionary change occurred, they suggested, only in short, rapid bursts, punctuated by long periods of stasis. The theory became known as punctuated equilibrium. Other biologists continue to insist that evolution is typically gradual and continuous, and the structure of the fossil record is merely a reflection of the unreliability of the processes that formed it.

Occasionally this last debate has been picked up by enemies of evolution as suggesting that evolution itself, the core propositions, remains a controversial thesis. The reality, however, is that this is a controversy entirely internal to the broad evolutionary programme. The same, indeed, can be said of all the issues I have just summarized. My aim in this chapter has been to show how these debates fit unproblematically with the core claims of evolutionary theory, and certainly offer no threat to any of these claims. That extant living forms are derived from earlier, often simpler forms, and that all living forms are related by descent, is as much an established fact as anything in science can ever be. But these ideas are not part of a frozen dogma. They are at the centre of a vigorous scientific research programme in which a great many issues are subject to lively debate, and no doubt others will become subjects of debate in the future. So, to summarize the discussion so far, there is a core to the theory of evolution which is simple fact, but evolutionary biology is a field of many vying theories. In between the undisputed facts of descent and relatedness, and the controversies over pace of evolution or units of selection, there are central ideas, most importantly natural selection, the importance of which is not in any doubt, though once again the details are the subject of continuing debate.

3

What is the Theory of Evolution Good For?

The fruits of science

The main object of this book, as noted at the outset, is not to question the value of the theory of evolution, but rather to pose the question, what does this undeniably exemplary scientific construct do for us? Why should we care about it? My thesis in this chapter is that while evolutionary theory provides us with unprecedented insights into the overarching narrative of the history of life, and enables us to see how many disparate sets of facts fit together, its ability to provide detailed explanations of specific phenomena is often greatly overstated. Evolution tells us a great deal about our place in the universe, but not nearly as much as is often supposed about the details of what kinds of beings we are. To understand the limits, and even the bankruptcy, of so much fashionable contemporary mythology about human evolution, it is important to have an idea about how evolutionary explanations really work.

We might begin an exploration of the value of evolutionary

thinking by asking what benefits we normally hope to derive from the products of any scientific activity. A natural next step is to distinguish two kinds of benefit, practical and intellectual. On the practical side, I have in mind two things frequently taken to be central excellences of science, prediction and control. Medical science, for instance, aims to predict the outcomes of various physiological interventions and insults to the body, and aims to use this predictive ability to maintain the body as near as possible to desired states of health. Evolution, by contrast, can be seen as a body of science with almost entirely intellectual benefits. There have been strands in eugenic thinking that have based their recommendations on speculations about the future of human evolution, but these have been widely discredited. The reason that evolution is not a practical science is obvious enough: most of its claims relate to periods of time far longer than those that are of any direct relevance to human life. Similarly, astronomy lacks much application because the spatial dimensions of its subject matter lie outside practical human concerns. There are, it is true, evolutionary processes that happen much more quickly, especially in micro-organisms. This is very important for parts of medical science, as it lies behind the discouraging facility of pathological organisms in acquiring immunities to drugs. However, important though this problem undoubtedly is, the central thrust of evolutionary investigation is directed at the very long term, and therefore at a timescale beyond our capacities for effective intervention.

Explanation

When we consider the intellectual rewards of evolutionary theory, two important ideas that figure largely in philosophical discussions of science are explanation and understanding. The two are closely connected. To explain something is to understand why, or at least some part of why, it happened. These two concepts have tended to come apart in the philosophy of science, however, because scientific explanation has been closely associated with a specific model, subsumption under universal laws. On this model, we explain an event when we can specify laws of nature and particular conditions that together imply that the event would happen. For example, the law of gravitation, together with the facts that my pen is an object heavier than air and has been released in the vicinity of the surface of the Earth, a massive object, entail that the pen falls to the ground. This is then held to explain the fact that my pen falls to the ground. Historically, this tradition in the theory of explanation is closely linked with the rejection of the possibility of getting any deeper insight into natural processes than the mere discovery of regularities. It has been supposed that the theory of gravitation, for instance, can do no more than summarize our extensive experience of falling objects and suchlike, and cannot provide insight into why natural regularities are as they are. So there is no more to explaining a phenomenon than showing it to be an instance of a kind of sequence that is observed to occur universally. David Hume, famously, offered highly influential arguments to the effect that there could be no deeper explanation of why nature was characterized by certain regularities. It is a common complaint that on this account scientific explanation doesn't give much understanding.

Recently the foregoing account of explanation has been subject to a good deal of criticism. For present purposes, all that needs to be said is that even if this is a good model for some kinds of scientific explanation, it is not adequate for all. And one area in which it is almost surely not much use is for evolutionary explanation. In fact, there seems little likelihood that evolutionary laws can be formulated at all. Recall the giraffes. There is surely no law that decrees that a population of tree-grazing animals will evolve longer necks. Being taller has structural costs and it is very likely physically impossible to acquire a neck long enough to graze in the forest canopy. There may well be limits to the possible height consistent with the mammalian body plan—this is an example of one kind of constraint on selection. The exact balance of costs and benefits will depend on what kinds of trees there are, how common they are, what competitors graze on what parts of them, and so on. And there is no guarantee that a population will produce the right mutations to generate a fitness-enhancing change, or even that there are any such mutations.

And evolution has irreducible elements of luck. Perhaps all the giraffes except a few fortunate ones that happen to be unusually short are wiped out by a freak storm, for example. The general moral is that evolutionary history, like human history, while sometimes highly intelligible, is nevertheless

also shot through with contingency. It is therefore most
unlikely that the evolution of the giraffe's neck instantiates
some universal generalization about the trajectory of populations of tree-grazing animals. The amount of information
one would need to assemble in order to infer reliably that
the giraffes must have evolved long necks, that any population of animals similar in the relevant respects were bound
to evolve comparably elongated supports for their heads,
would be so great as to limit the application of the explanation to the particular case in point, and thus rule out any
relevance of, or evidence for, a subsuming generalization.

Despite failing to meet the rigorous demands of the
inferential theory, evolutionary explanations may also be
seen as offering more understanding than inference from
law will normally provide. This is because evolutionary
explanation seems, at least, to provide an answer to the
question of what some feature of an organism is for. If the
nursery story of giraffe evolution is true, it is tempting to say
that reaching high leaves on trees is what the giraffe's long
neck is for. It has become orthodox among philosophers of
biology to propose that this is just what it means to say that a
feature of an organism has a particular function: the reason
it is there is because it provided greater fitness to ancestral
organisms, and these were therefore selected. The function
of the eye is to see if, and only if, creatures have eyes because
ancestors that could see better were selected over those that
could see less well. This is a good example to make the
point, since it is beyond dispute that seeing is the function of
an eye, and for the same intuitive reasons beyond dispute

that selection for the ability to see must be central to the explanation of organisms coming to have eyes.

The complexity of evolution

I think the association of function and evolutionary history just described is a mistake, however. This is because, in the end, I am sceptical about the capacity of evolutionary considerations to provide the kinds of explanations just sketched, explanations of the presence of particular features of particular kinds of organism: that giraffes have long necks or that peacocks have gaudy tails. Now of course, in such cases, there is some increase of understanding when we realize that long necks enabled ancestral giraffes to reach higher leaves on trees, or peacocks with flashier tails attracted the interest of ancestral peahens. It is not at all my intention to question the importance of the processes of natural selection or sexual selection. What I do want to suggest, however, is that it is not at all easy to fit these cases into any familiar model of explanation.

So what is wrong with these familiar explanations of necks and tails? The quick answer is just the complexity of the evolutionary process, but it is too easy and too common nowadays to wave vaguely in the direction of complexity, and certainly more needs to be said. We can begin to see the problem if we focus very clearly on what we take there to be in the world that corresponds to the stories we tell about natural selection. In the end there is only a sequence of organisms, being born, growing, reproducing (or not), and dying. These salient events have causes, and we suppose that when we pick out features that we suppose were selected we identify factors that have tended to cause reproduction, prevent death, and so on. But is this enough to make sense of separating the selection of the peacock's tail from the selection of peacocks, or the giraffe's neck from the giraffes? My suggestion is that it makes only very variable and limited sense. To see this, we can focus first on the question of where we see the beginning of such subordinate processes, and second how we see them as separated from everything else that is happening simultaneously.

The first difficulty can be approached through a technical distinction that is sometimes made between adaptation and exaptation. A simple-minded view of this distinction begins with the idea that typically traits of organisms evolved because they served some function that they can be seen now to promote. These are called adaptations. However, sometimes the organism puts the trait to a use that is different from that which explained its selection, and that has come to be known as a case of exaptation. A classic example is the evolution of lungs from swim-bladders. Mammalian lungs, it is generally supposed, developed gradually out of the organs that ancestral fish used to keep themselves afloat; therefore they are exaptations. But in reality exaptation isn't just an anomaly that occasionally derails attempts at adaptationist explanation, it is a feature of almost any interesting evolutionary explanation. Organisms evolve within a range of possible trajectories that is, from the point of view of the entire space of biological possibility, very narrow. The resources with which they have to work are what they are because of previous evolutionary processes, and to evolve an organ for doing X one must usually start with an organ or structure that evolved to do Y. The giraffe and the peacock, to continue with our stock examples, already had a neck and a tail when they started eating high leaves on trees or when the peahens started to take an interest in flashy tails. And presumably these necks and tails already had their own selective histories. Many of Stephen Jay Gould's wonderful essays in natural history illustrate this aspect of evolution. One example he has made famous is the panda's thumb, cobbled together, *faute de mieux*, out of fairly unsuitable bits of wrist bone.

The second and more obvious point is that only in the simplest possible cases is the relevant evolutionary history simply a matter of selection on the basis of some particular adaptive consequence of a particular trait. Organisms are highly integrated systems, and changes to one trait will cause correlative changes to other traits and these will have positive and negative effects on fitness. Even the trait that is the primary focus of attention will typically have many fitness effects. A wing is fairly obviously for flying, but it may also scare off predators, shelter offspring, or get caught in the brambles. And the giraffe's neck, apart from some rather basic functions like attaching the head to the torso, is surely a structural liability.

In summary, then, if we try to imagine the actual causal process of evolution of, say, the giraffe's neck, all we have is a long sequence of populations of animals and a secular trend in neck length. Both the current neck and its physiological precursors will have been doing lots of things, some good and some not so good, for their proud owners. Selection for neck length is not readily extracted from this scenario. One obvious move here is to resort to a much weaker view of explanation. Let us suppose that, other things being equal, most of the time, having a longer neck caused ancestral giraffes to have a higher probability of surviving and reproducing. Then we have identified a factor that raised the probability of the effect under consideration. A number of philosophers now consider that for a factor to be explanatory of a particular effect is just for the presence of that factor to make the effect more likely.

One difficulty with the idea just mooted is a familiar one in discussions of evolution, the threat of vacuity. The phenomenon we originally wanted to explain was an historical trend towards greater neck length. We therefore know from the outset that longer-necked animals tended to survive and reproduce more successfully than their shorter-necked fellows. Have we added much to this story when we add that they must have had a higher propensity to survive and reproduce? The answer, I suppose, is that we have gained understanding to the extent that we understand why possessors of the feature of interest proved fitter. And now the problems described earlier, of exaptation and of multiple and interconnected effects, suggest that we may typically be unable to provide more than a few tentative suggestions on this point and certainly not more than a fragment of the truth.

Perhaps this is the wrong sort of case. It would be difficult to establish that the fitness advantages of longer necks were invariably a consequence of access to higher leaves and eliminate a host of other possible advantages or even a purely allometric response to some other trend. ('Allometry' refers to the developmental correlations that require that when one feature changes, others may do so purely to accommodate the first change. Bigger animals have bigger brains, for instance.) But, to move to another classic case, there can be no doubt that the eye is designed for seeing. The design of an eye could only have come about, we suppose, if organisms with better functioning eyes had proved fitter than those with less visual perspicuity. And surely this is right.

My only point here is that we are back to very little more than the general metaphysical commitment discussed in the previous chapter. We have a general explanation of the possibility of adaptation, or design, in the theory of natural selection. We are confident that a biological trait as clearly adapted as the mammalian eye must owe its existence to natural selection, and hence we believe that a complete genealogy of, let us say, the vertebrates, will disclose a trend towards more complex and efficient visual perception systems. And we suppose that the advantages to their possessors of more efficient eyes had some positive influence on their survival. And perhaps this is sufficient to count as an explanation, if not an exceptionally illuminating one.

It is worth mentioning here that many philosophers, and most especially those concerned with biology, have abandoned altogether the conception of science as a search for universal generalizations or laws, and have focused instead on the production of models. Scientific models are generally conceived as providing only one perspective on the phenomena under study, or one aspect of what is happening in reality. Thus one might provide a model in which a population of organisms differ only in their visual acuity, and consider the evolutionary trajectory of such a population. But, of course, no real populations are that homogeneous, so it is obvious that the relationship between such a model and reality is a tenuous one, and certainly no single model can claim to provide the canonical representation of an actual population. Good models may nevertheless be seen as providing significant, though partial, insight into complex processes.

The limits of evolutionary explanation

Let us assume now that we have arrived at a modest but not entirely vacuous conception of evolutionary explanation. If we are sufficiently confident of the basis for the fitness advantage of a feature to have no doubt that those advantages played an important role in the origin of the feature, then identification of those advantages provides us some kind of evolutionary explanation of the feature's presence. I have two modestly sceptical comments about such explanations. My scepticism is not intended to undermine the legitimacy of such explanations, but rather to show that they require quite special circumstances to function. The important conclusion is that we must not assume that any feature of an organism that attracts our attention is an appropriate subject for evolutionary explanation. This caution is an essential preliminary for seeing how flimsy are the underlying assumptions of so much contemporary speculation about human evolution.

My first sceptical comment is that the possibility of even such broad explanation sketches is limited to those cases where the function of the trait in question is beyond any serious doubt. This is true of the eye, I suppose, but probably not of the giraffe's neck. A neck is a fairly basic structural component of a vertebrate and is used for many things. It is likely, for example, that the advantages of reaching higher leaves must sometimes be outweighed by the disadvantages of trying to reach down for ground-level vegetation, not to mention the general nuisance of carrying the thing about. And there will surely be a range of allometric consequences of extending the neck. Does this mean that there is no explanation for the giraffe's neck? My suggestion was that nothing much less than a full history of the lineage complete with varying ecological and climatic circumstances, not to mention brute chance, is likely to provide such. Whereas the eye has an unmistakable function, the neck has many functions and is structurally interdependent with many other parts of the organism. There is no simple selective story to tell about it. We should recognize that many parts of organisms are like this.

This suggests a simpler way of getting to the same point. I

have talked a good deal about explaining the features of organisms. Evolutionists often refer to such features as traits. But there is no naturally given way of dividing an organism into features or traits. (Compare the question how many things there are in a room. A table? A table top and four table legs? Is that one thing, five, or six?) There are many reasons why we might be interested in particular aspects of organisms and wonder about the evolutionary origin of these aspects. But evolution doesn't see the trait, it only sees the organism. There should be no surprise, then, that there is no right story about how evolution selected the trait. To anticipate a later chapter, the division of human behaviour into discrete traits looks even more unpromising than the similar atomization of an organism.

I have concentrated on one particular direction in which one might look for explanatory benefits from evolutionary theory, the project of explaining the presence of particular features of particular kinds of organisms. I have concentrated on this because it is the area in which the fruits of evolutionary theory have been most controversial, and also because it will set the scene for the important arguments in a later chapter concerning evolution and human nature. To avoid serious misunderstanding I should reiterate that there are other things that evolution explains much less controversially. Here I have in mind the evidence summarized in the previous chapter for the truth of evolution, the facts that articulate the details of the central theses of the theory of evolution. These are the facts of comparative morphology, of geographical distribution of similar forms, and of the positions of objects in the fossil record. The consilience of these explanations—their co-explanation by the same ultimately quite simple set of ideas—provides the case for the truth of evolution but, at the same time and by the same token, provides domains of phenomena explained by the theory. The diversity of these domains blocks the suggestion that this collocation of evidence and theory is viciously circular. Further exploration of these domains allows further articulation of the theory and also of the facts of evolutionary history. One project within evolutionary studies is the production of increasingly accurate knowledge of the actual sequence of living forms that constitutes the genealogy of terrestrial life. This involves further accumulation of the kind of evidence that the theory explains, and facilitates knowledge of the particular conditions under which the theory has operated. Such ongoing research continually improves the coordination between theory and evidence. My scepticism about particular applications of evolutionary theory in no way threatens the evidential support and scientific respectability of the theory in its proper place.

In the next chapter I turn from the strictly scientific interpretation of evolution and its consequences to what might be called its metaphysical implications, the effect that the establishment of evolution as fact has on our general view of the world and our place in it. Contrary to a line of argument that has been in much evidence recently, especially in the US, I consider this perhaps the greatest significance of Darwin's work.

4

Human Origins and the Decline of Theism

Naturalism

The core propositions of evolutionary theory are not much open to debate. They also have profound implications for our understanding of our place in the world. I shall argue in this chapter that Darwin's theory provides the last major piece in the articulation of a fully naturalistic world-view and hence would, if fully appreciated, deliver a death blow to pre-scientific, theocentric cosmologies. It is a commonplace that the growth of science in the last several centuries has gradually eroded the grounds for religious belief, and that Darwin's contribution was of particular significance to this process. Like most commonplaces, it is highly simplistic. Darwin's own relation to religious belief is complex, and it was almost certainly not among his intellectual objectives to overthrow Christianity. However, my own interests here are not historical but philosophical, and I want to claim that whatever Darwin's goals, and whatever his contemporaries may have made of his ideas, the growth of evolutionary theory that he launched has provided a fatal injury to the pretensions of religion.

I described my own perspective as 'naturalistic', and I should explain what I mean by this rather slippery word. This word has many uses in philosophy, and some of them are quite controversial. Here I mean something relatively simple, something that might more perspicuously, if inelegantly, be described as anti-supernaturalistic. As this term is intended to suggest, my naturalism refuses to countenance the existence of ghosts, souls, and, of special importance here, gods. This may seem intolerant or arrogant: surely we don't know everything that exists and there may well turn out to be things very different from those currently countenanced by respectable scientific academies. This I happily acknowledge. But my objection to souls, deities, and the like is not mere prejudice. It is based on a principle, the principle that our belief in the existence of things should be grounded, ultimately, on experience. This is the principle of empiricism that has been central to much of Western philosophy for several centuries or perhaps millennia. The principle must be interpreted broadly: the experience on which we base our belief in electrons, say, is complex, and not simply related to the presence of electrons. Similarly with black holes. Nonetheless, there is relevant evidence and there are arguments by which the evidence is connected to the conclusions it is held to support.

There are philosophical issues here I shall avoid. There is a sense, no doubt, in which numbers and perhaps a great range of abstract objects may truly be said to exist.

Philosophers debate the existence of their own esoteric theoretical entities, such as universals, curious entities designed to explain how different things can share the same property. But no one thinks these things exist in space and time in the way that more mundane and concrete entities do. And it is in this sense, as part of the 'furniture of the world', that I want to consider the existence of ghosts, souls, and suchlike. God is sometimes said to be outside time. I don't pretend to understand what that means, but I take it he is, at any rate, considered on occasion to intervene in space and time—by creating the universe, for instance, or by sending his son (or himself) into the world to redeem our sins. And that is enough for me. An entity that intervenes in space and time can, while so intervening, provide empirical evidence for its existence. If it fails to do so, I remain so far unconvinced of its existence. Of course, God, when he is said to have spent time on Earth, was empirically perfectly accessible. The trouble is that he was, at the time, by all accounts, a man. The suggestion that he was also a deity was not so open to empirical evidence, and so the claim that there are deities at all remains devoid of evidence.

Arguments for theism

I don't propose to say much more about the basis for my commitment to naturalism (or anyhow anti-supernaturalism). But I will say something about its application to the particular case of theism. There is a venerable tradition of arguments for the existence of God that purport to be a priori, to be independent, that is to say, of how things in the world appear to our senses. The most famous such argument, known as the ontological argument, attempts to show that the concept of God is one such that, necessarily, there must be something to which it applies. Put very simply, this argument begins with the premiss that God is, by definition, perfect. Since non-existent things are less perfect than existent things, God to be perfect must exist. Therefore God, by definition, exists. I am glad to say that this argument carries little conviction these days, and it is something of an embarrassment attempting to persuade first-year philosophy students to take it seriously enough to work out exactly where it goes wrong.

Any such argument is ruled out by a quite different philosophical tradition, for which the great Scottish philosopher David Hume was perhaps the most articulate spokesman, the empiricist tradition that denies that anything substantial about the world can be known apart from experience. Since the existence of God would be a substantial fact, these traditions contradict one another. The combination of empiricism and naturalism I espouse falls within this latter, Humean, tradition. However, quite apart from this commitment, a priori arguments for the existence of God are nowadays acknowledged, even by most theologians, to be largely without merit.

Contemporary theologians, however, do not on the whole conclude from the failure of a priori arguments for God's existence the necessity for providing empirical ones. Rather, they have tended to give up the commitment to arguments altogether and, following some suggestions attributed to Jesus, appeal rather to faith. Faith is, I suppose, a rough synonym for belief, with an additional connotation that this belief is not grounded on anything. This is a difficult concept to take seriously from a philosophical point of view. Obviously if one has it, one finds it convincing. If one doesn't, it's hard to know how to understand the conviction, since no reasons can be offered for it. The difficulty is exacerbated by the variety of objects of faith that are on offer. How does one decide whether to be more impressed by the convinced Christian or the convinced Muslim? Or, for that matter, the person equally convinced of the healing powers of crystals or that faith can move mountains? If there are no reasons for adopting these systems of belief, it seems impossible for there to be any reason for choosing between them. My own response is, I need hardly say, not to take faith very seriously. It is one thing to admit that our knowledge of the universe is extremely limited, but a counsel of despair to respond to this by believing whatever we feel like. It goes without saying that arguments of this sort will seem quite beside the point to the faithful believer (though, as a matter of fact, I do have some scepticism as to whether faith is really as impervious to reason as the concept suggests).

It may perhaps be a matter of definition that some things, were they to exist, would existence. They may, that is to say, be defined as beyond normal sensory contact. I have no interest in such putative things. As already implied, I do not take God to be such a thing. I take it that God, should he choose, would have no difficulty in making his existence empirically unmistakable. He could appear on Earth in some suitably impressive guise and perform all manner of miracles. He might, for instance, institute world peace. No doubt theologians have many good explanations for his decision to eschew this kind of behaviour. At any rate, the question that concerns me is whether the reasons actually offered for the belief in various entities is empirical, in my broad sense, or not. The main point I want to make in this chapter is that prior to the development of a convincing theory of evolution there was an argument of sorts for belief in God, and an argument that could have been seen to meet naturalistic standards. However, this argument, always problematic, was entirely undermined by the development of a convincing account of evolution. Consequently, I claim, we now have no good reason for belief in God. This is, of course, a very major contribution to our world-view.

I should mention the possibility that there are moral rather than empirical reasons that favour religious belief. It is, of course, enormously problematic to offer as a sufficient reason for belief the suggestion that one would be better off believing it. This is generally described as wishful thinking. The great American philosopher and psychologist William James made an impressive attempt to defend such a procedure, but without convincing many. At any rate, though I shall not attempt to argue this notoriously controversial point in any detail, I am personally extremely sceptical of the thesis that religious belief is generally conducive to human welfare. A quick reflection on current events suggests that a great many major world conflicts involve a divide between people with major religious differences. Of course it will be said, no doubt correctly, that the ultimate causes of conflict are not outrage at perceived theological errors, but much more mundane economic and political rivalries. Nonetheless, criteria are needed for dividing people into groups deemed worthy of favourable or unfavourable economic or social treatment. Historically, nationality or race have been sufficient for this purpose, but the recognition of the superficiality of these tribal divisions is increasingly obvious to even mildly progressive thought. Religious difference, arguably, remains the most effective basis for defending boundaries between them and us, and the withering away of this kind of mythology would, I think, be entirely salutary. This is to say nothing of the thought that it may well be better for people to believe what is true.

The argument from design

The main task of this chapter is to consider whether there are any prospects for finding empirical arguments for the existence of God, in the absence at least of God deciding to disclose his existence empirically. The classic argument that aims to offer evidence for belief in some deity or other is the so-called Argument from Design. In simplest outline this argument says that the world, or some of the things in it, show unmistakable marks of design; therefore there must be a designer; that is to say, there must be a God. The argument is an ancient one, perhaps suggested by St Paul (Romans 1: 20): 'For since the creation of the world God's invisible qualities—his eternal power and divine nature—have been clearly seen, being understood from what has been made, so that men are without excuse.' The most famous exponent of the argument was William Paley around the beginning of the nineteenth century, and Paley's best-known figure is the comparison of nature and a watch:

> In crossing a heath, suppose I pitched my foot against a stone and were asked how the stone came to be there, I might possibly answer that for anything I knew to the contrary it had lain there forever; nor would it, perhaps, be very easy to show the absurdity of this answer. But suppose I had found a watch upon the ground, and it should be inquired how the watch happened to be in that place, I should hardly think of the answer which I had before given, that for anything I knew the watch might have always been there.

I should, of course, conclude that someone had designed and constructed the watch. Paley then points out that elaborate though the workings of a watch may be, they are simple compared to those of a plant or animal. Consequently the argument for a designer in the latter case is stronger even than that in the former. Hence nature must have been designed and created, and this designer and creator we refer to as God. This argument was required reading for English university students for much of the nineteenth century, and one student who was impressed by it was Charles Darwin:

In order to pass the B.A. examination, it was, also, necessary to get up Paley's *Evidences of Christianity*, and his *Moral Philosophy* ... The logic of this book and as I may add of his *Natural Theology* gave me as much delight as did Euclid. The careful study of these works, without attempting to learn any part by rote, was the only part of the Academical Course which, as I then felt and as I still believe, was of the least use to me in the education of my mind. I did not at that time trouble myself about Paley's premises; and taking these on trust I was charmed and convinced of the long line of argumentation.

(Charles Darwin, *Autobiography*)

The argument has a good deal of superficial plausibility, and it has convinced many. And, as I have also noted, it is an argument that starts from the evidence of the senses. Its merits do not, however, survive close scrutiny, something to which it was subjected a little before Paley wrote his *Natural Theology* by David Hume in his *Dialogues Concerning Natural Religion*. Hume's rebuttal of the argument from design is devastating. A first problem to note with any version of the argument is that even if it succeeds in showing that the world must have a designer, it has little power to disclose what that designer is like. Hume distinguishes sharply the question of whether natural theology (the general project of inferring theology from examination of the world) can show that there must be a designer from the question of whether we can attribute to this designer the moral properties traditionally associated with God, and his arguments against the latter project are even more devastating than those against the former.

The first and longest part of Hume's dialogues addresses the first of these questions. The general form of the argument from design is one that has received considerable discussion in recent philosophy under the title 'inference to the best explanation'. As this name suggests, the general idea is that we can infer the truth of some, typically unobserved or even unobservable, fact from the recognition that that fact would provide the best explanation of some observed fact. This is a form of argument sometimes held to be of great importance to scientific thinking because it appears to be required if we are to have grounds for believing in objects such as electrons or quarks that are not accessible to direct observation. Physicists may speak of observing an electron when they see a track in a bubble chamber, but all they see is a line of bubbles in a tank of liquid. The electron, it is suggested, is inferred as the best explanation of the occurrence of the track. The argument is controversial, but let us assume that it is in principle sound. Or, at least, that some instances of it are sound.

One thing sound instances of inference to the best explanation surely require is that there be one explanation identified as the best. This is a main point at which Hume (or at any rate his spokesman in the dialogue, Philo) criticizes the argument from design. Allow, says Hume, that someone has succeeded in establishing that there was, indeed, a designer of the universe. Even then:

Beyond that position he cannot ascertain one single circumstance, and is left afterwards to fix every point of his theology by the utmost license of fancy and hypothesis. The world, for all he knows, is very faulty and imperfect, compared to a superior standard; and was only the first rude essay of some infant deity who afterwards abandoned it ashamed of his lame performance: It is the work only of some dependent, inferior deity, and is the object of derision to his superiors: It is the production of old age and dotage in some superannuated deity . . .

(*Dialogues*, part V)

In fact Hume is not always willing to accept the assumption of any designer at all. At one point his characters toy with the idea that the universe is more like an organism, and perhaps was the result of quasi-biological reproduction. On the whole some super-vegetable seems the most promising analogy. Of course, this invites the question where the parental universes came from; but it is equally a worry with the argument from design that if the existence of an ordered object requires the hypothesis of a designer, it is hard to see why God himself, presumably being supremely ordered, does not require a designer. Most interestingly of all, he even flirts with an idea that foreshadows something of Darwin:

Is there a system, an order, an economy of things, by which matter can preserve that perpetual agitation which seems essential to it, and yet maintain a constancy in the forms which it produces? There certainly is such an economy, for this is actually the case with the present world. The continual motion of matter, therefore, in less than infinite transpositions, must produce this economy or order, and, by its very nature, that order, when once established, supports itself for many ages if not to eternity.

(*Dialogues*, part VIII)

Richard Dawkins, in *The Selfish Gene*, noted that the survival of the fittest is a special case of a more general and self-evident principle of the survival of the stable. In infinite time, Hume suggests, the universe will find by chance the kinds of stability we know, from their actuality, to be possible, and as a result of their stability those states will endure.

Hume is quicker and more uncompromising with the idea that we might infer the existence of a creator with the benevolence usually attributed to the Christian God. The world is filled with avoidable evil—pain, disease, natural disasters, and so on—and the creator of the universe either lacked the power or the will to prevent these evils. But of course an omnipotent being could not lack the power, and an infinitely benevolent one could not lack the will, so natural theology contradicts some of the traditional attributes of the deity. Philosophers and theologians have gone to great trouble through the ages to establish that the nature of the world is not wholly inconsistent with a benevolent creator (it's not really as bad as it looks; evil has distant good consequences we're not aware of; and so on). But the prospects for showing that the universe is so full of goodness that it could have been created only by a supremely benevolent being are about as unpromising as for any argument one could come up with.

I won't dwell on these latter worries, however, because the earlier reflections on the grounds for believing in any God at all have already shown how little natural theology can possibly tell us about the nature of the creator. The importance of the second set of arguments arises if one accepts that some sort of designer is necessary, but supposes erroneously that this leads us to something resembling a traditional conception of God. As a matter of fact, despite the force of his own arguments, Hume does appear, at the conclusion of the *Dialogues,* to accept that the universe does provide evidence for a cause of order which will 'probably bear some remote analogy to human intelligence'—not exactly a triumphant vindication of theism, but not a complete rejection either. There is disagreement as to whether Hume meant even this much theism seriously or whether it merely reflects a sensitivity to the limits of acceptable speculation in the eighteenth century. But we need not worry about this question here, because we can now see, as Hume could not, how the theory of evolution changes the entire argumentative context we have been reviewing.

I noted that it was essential to an inference to the best explanation that there be a clear candidate for the best explanation. And we saw that part of the weakness of the traditional argument from design is that there isn't a unique explanation, and certainly not a uniquely best one. And that, finally, is what evolution provides us with: a clearly and uniquely best available account of the origin of terrestrial life. I do not need, at this point, to insist that evolution provides the best possible explanation. It may be that one day we will find a much better one. But to render the argu-

ment from design harmless, we have only to note

that evolution provides a vastly better explanation than does a wholly vague appeal to a creator about whom nothing whatever is known.

Why do I insist that evolution is a much better explanation than is an intelligent designer? Because evolution is a theory that provides a detailed fit between explanatory apparatus and phenomena. The observations adduced in support of evolution earlier—homologous structures in different kinds of organisms, geographical relations between related species, the arrangement of fossils in the geological record, and so on—all form a rich pattern that is systematically articulated with the theory that in part explains them all. As was clear from Hume's criticisms, far from providing a comparably rich and detailed structure, the argument from design only specifies the theoretical apparatus in terms so vague as to border on meaninglessness, and explains only the presence of some order or structure, not a single detail of the actual structure found in our world. There is no possible comparison in richness of explanatory accomplishment.

The preceding conclusion as to the respective merits of evolution and natural theology is largely uncontroversial among those who have thought about the matter, notwithstanding the continuing opposition of creationists, including a very small subset of thoughtful creationists. It would not be true to say the argument has been abandoned. Indeed, it has recently re-emerged under a slightly different guise as Intelligent Design theory. However, as I am unable to see any genuine novelty in this approach, I shall not discuss it here. Curiously, however, the general consensus on the explanatory superiority of evolution over theism has not generally been recognized as constituting a problem for religious belief. Indeed some rather surprising people have devoted considerable effort to denying that evolution provides any threat to religion.

Attempts to reconcile evolution and theology

Two prominent evolutionists, the biologist Stephen Jay Gould and the philosopher and historian of biology Michael Ruse, have recently published book-length defences of the compatibility of Darwinism and Christianity. It is perhaps worth mentioning that both these authors were writing in the US, and that the political power of Christianity in that country is such as to make it highly expedient to minimize possible conflicts between science and religion. However, since opposition to evolution comes almost exclusively from fundamentalist Christians, and certainly neither Ruse nor Gould suggested that evolutionary theory was consistent with a literal interpretation of Genesis, this expediency is limited. I shall, anyhow, assume that both are entirely sincere. It seems to me that both are, in that case, misguided.

Ruse considers in some detail the main points at which Darwinian and Christian beliefs are alleged to contradict one another, and concludes that there are no such contradictions and indeed, for example in relation to Christian and sociobiological perceptions of ethics, they are sometimes mutually supporting. Often, of course, this is because they are mutually irrelevant. Darwin has nothing to say about the Trinity (as far as I know) and, while his work has little room for virgin births or resurrections from the dead, there is no need for him specifically to deny that a sufficiently powerful supernatural being could pull such things off. So I do not at all want to argue that there is a contradiction between Darwinism and Christianity. However, and here is the point noted at the beginning of the book where I find myself in disconcerting accord with Richard Dawkins and Christian fundamentalists, it seems to me that the argument misses the crucial point. This is the point argued in some detail in this chapter, that Darwinism undermines the only remotely plausible reason for believing in the existence of God. And, some extreme liberal versions of Christianity apart, belief in the existence of God does seem to be a minimal condition for Christianity. Consequently, and contrary to the orthodox philosophical view of the matter, I believe that Christians—not merely fundamentalist Christians—are quite right to try to undermine Darwinism, and Richard Dawkins is quite right that, since their attempts to do this are wholly unsuccessful, there is nothing worthwhile left of the argument from design. More contentiously, I want to insist that without the argument from design there is nothing very credible left of theism generally, and Christianity in particular. Hence Ruse's argument for compatibility, while generally successful, seems to me largely beside the point.

Gould achieves the conclusion of compatibility in a more drastic way than does Ruse, by insisting on a radical division between two spheres of thought or 'magisteria'. One, the magisterium of science, is concerned with how things are, while the other, the magisterium of religion, is concerned with ethics and, in the traditional sense so distressing to professional philosophers—and especially philosophers of biology—the meaning of life. These are wholly independent domains, and therefore no possible threat to one another. Gould convincingly demonstrates that many of the greatest scientists and theologians have recognized some such distinction and often been deeply committed to both magisteria. As science has advanced the cannier theologians have quickly withdrawn their claim to factual authority in the domains that it colonized, and retreated increasingly to the distinct magisterium of ethics and the meaning of life.

Despite the impressive authority that Gould marshals for this position I am quite unconvinced, for two main reasons. The first, and perhaps most fundamental, is the point that I have stressed throughout this chapter, and specifically in connection with Ruse's thesis. It is that science, especially in the guise of Darwinism, has undermined any plausible grounds for believing that there are any gods or other supernatural beings. Indeed, this simple observation, whether or not it is true, points to the difficulty of keeping Gould's magisteria neatly separated. For whether or not there are any gods is a pretty fundamental question about how things are. Science doesn't declare that there are no gods, but it certainly doesn't say there are any. So if science were really the only magisterium licensed to say how things are, we would have no reason for saying there was a God, and the magisterium of religion would be pretty seriously flawed.

It is no doubt true, on the other hand, that many religious

traditions have provided us with deep and cogent thoughts about how we should live and provide our lives with meanings. We may greatly respect these traditions and their advice without supposing that their value has anything to do with a supernatural origin. Indeed, even if they did have a supernatural origin, there is a familiar and compelling concern that unless there were independent reasons for judging them to be valuable—reasons presumably known to God—it is unclear why we should act on them, unless perhaps out of fear of recrimination. And if there are sufficient independent reasons God's endorsement becomes superfluous. My point is that even if Gould is right that these two realms of thought are quite distinct, this does nothing to save the theological claims—centrally the existence of a God—from the threat that they encounter from the growth of science. So it seems to me entirely misleading to present his argument as addressing the relation of science to religion. Put another way, Gould does nothing and attempts nothing to show that there is a basis for ontological commitment wholly independent of science (or perhaps everyday experience). And I very much doubt whether he would, if pushed, have been interested in doing so.

My second worry about Gould's argument is that I doubt whether he is right to think that a sharp line can be drawn even between science and ethics. There is a long tradition in philosophy of assuming such a divide, something that has become known as the fact/value distinction, but which has become increasingly controversial in recent philosophy of science. At any rate, Gould assumes a rather old-fashioned

sounding version of the fact/value distinction. But one of
the areas where the fact/value distinction comes under most
obvious pressure is precisely in the context of evolution.
Evolutionists, not always credibly it is true, make claims
about what we must fundamentally be like grounded in their
understanding of evolution. But it is surely likely that defensible claims about how we should live would be very different
if we were either solitary, asocial creatures or, at the other
extreme, social insects. If it were demonstrated, for instance,
that all properly developed humans had an obsessive urge to
accumulate personal possessions, the proposal that humans
were morally obligated to live in communist societies with
no personal possessions whatever would be very difficult to
defend. The curious blend of individualism and sociality that
in fact characterizes *Homo sapiens* is no doubt what makes
the question of how humans should live so difficult and
interesting. (Ironically, sociobiologists and their intellectual
descendants often make parallel claims to a rigid distinction
between facts and values, in order to insist that their claims
about human nature have no relevance to ethical or political questions.) Presumably again, if we were partially supernatural creatures made in the image of an omnipotent being,
this would also have consequences for the nature of a good
human life. One commonly held ideal of a possible good
life is one spent in adoration of or service to the Supreme
Being. It is hard to believe that the value of such a life is
independent of whether there is, in fact, any such being to
adore or serve. In sum, how we should live is a question that
cannot be wholly separated from facts about how things are.

The main point, which would perhaps be unnecessary to labour if it were not so controversial and if it had not been denied in important respects by some quite unlikely people, is that the theory of evolution has been a major, even decisive, contributor to the process of undermining prescientific supernaturalistic metaphysical views and replacing them with the naturalistic metaphysics assumed by most contemporary philosophers. The question is not whether evolution and a particular religious tradition are logically consistent. Provided the religious tradition avoids factual claims, as Gould's conception of distinct magisteria forces them to do by fiat and as sensible theologians have been increasingly willing to do for centuries, they are consistent because they do not speak on the same subjects. But it is nevertheless the case that science and religion speak for radically different conceptions of the universe. And as the conception fostered by the former has become more compelling, so that promoted by the latter has become less tenable. Science does not contradict religion; but it makes it increasingly improbable that religious discourse has any subject matter.

It is sometimes suggested that science will always leave gaps, and those gaps will provide congenial habitats for a residuum of religion. Certainly we can be confident that science will never succeed in explaining everything, and if it is felt useful to appeal to the heavens for explanations where science fails, then so be it. The trouble is that it is difficult to choose the right gaps: religion has tended to choose gaps (life, the mind) which science has subsequently begun to fill, and the retreat to safer ground is a poor advertisement for the explanatory virtues of religion. However, there is one rather large gap within the actual domain of evolution that must be mentioned. That is the very beginning of life.

It is true that the ultimate origins of life, the transition from primeval slime to the first living cell, remain little understood. Should one conclude that the explanation of this transition in terms of rough and tentative sketches of chemical processes is no better or worse than appeal to divine intervention? Hardly. First, recall that my concern has been whether inspection of the world in which we find ourselves gives us grounds for believing in a supreme being. If we had already established that there was such a being, and that he was a being with an aptitude for creating life, then his intervention might indeed be the best explanation for the beginnings of life. But we are now contemplating a being whose existence is to be postulated solely as an explanation for this puzzling phenomenon—an ad hoc manoeuvre of a kind that is generally agreed to vitiate any claim to explanation. On the other hand, while there is much that we do not know about the dawn of life, we are not without explanatory resource. We do know that the conditions that seem likely to have obtained a few billion years ago when life on Earth probably began were suited to the production of fairly complex organic chemicals. There is some reason to think that complex mixtures of organic chemicals may establish stable patterns of reaction not dissimilar to the metabolic dynamics found in biological cells. We have good reason to suppose that once minimal conditions of replication of something cell-like had been accomplished, natural selection might be expected to bring about increases in sophistication and stability. And there are many much more detailed speculations about the chemical details of these hypothetical processes.

The situation points to an important aspect of inference to the best explanation. The best explanation available may well be, and perhaps often is, sketchy and partial. We are reasonably confident that the movements of astronomical objects are best explained by well-known principles of mechanics even if there may be theoretical mathematical obstacles in the application of those principles to that case. I readily admit that in the present case the leap is longer. We don't have any exactly parallel cases to which to appeal: there are real gaps in our understanding of the processes at work. But, by contrast, appeal to a supreme being strikes me as being no explanation at all. We know nothing about how such a being creates life or anything else; we don't know why he creates one kind of life rather than any other; and, in short, we are admitting that the problem is beyond our powers, and imagining a being for whom this is not the case. The most profound implication of evolution is that it should finally make clear to us that we neither have nor need an all-powerful father figure to take on the tasks that seem presently beyond us. It may well be that there are things about the universe that we shall never understand. But there is no one else to do this for us—or at least no one but, perhaps, other naturally occurring and evolving occupiers of space and time in other parts of the universe.

5

Humans and Other Animals

Souls

Bishop Samuel Wilberforce's reluctance to accept that his ancestry might include an ape, even if Charles Darwin's did, is one of the better-known anecdotes in the history of science. It also points to another familiar reason why the religiously inclined have sometimes been hostile to Darwinian theories; namely, their fear that the fundamental divide between humans and the rest of creation will be undermined by the doctrine that all organisms share a common ancestry. And no doubt there is something to this. Continuity of descent does suggest limits to the depth of the divide that is imaginable between humans and their animal relatives. The object of this chapter is to explore the nature and extent of this divide. Although I wholly reject the absolute distinction between us and them still maintained by much religious thought, I shall propose a wider gulf than is commonly admitted by many contemporary scientists and philosophers.

A position which is indeed rendered problematic by evolution is the dualism famously propounded by the great French philosopher René Descartes, which distinguishes between humans and other animals on the basis that whereas the latter are mere mechanisms, the former possess immaterial minds. For present purposes it will do no harm to equate the Cartesian mind with the Christian soul. While it is not impossible to imagine a creator who watched the course of evolution and at some point decided (or, more likely, enacted an earlier decision) to ensoul a privileged subset of his creation, the picture seems implausible. In part, of course, it is implausible for one of the main reasons that most philosophers have abandoned Cartesian dualism. Whatever the unique features of humans may be, there are realms of behaviour for which the similarities between ourselves and our near relatives are too great for it to be credible that in one case the behaviour reflects an underlying soul or mind, while in the other there is no such thing, only the grinding away of neural machinery.

One response to this difficulty is to assume that, in essential respects, whatever amounts to the human soul is also to be found in other creatures. For instance, it is very difficult to believe that other animals feel no pain, given the similarity of their response to physical damage to our own, and even more difficult—probably unintelligible—to suppose that they may have pain, just pain of which they are not conscious. Recent philosophy has shown a revival in interest in the nature of consciousness. I have some doubts whether the problems central to this revival are well posed. But however they are posed and answered, if such answers are to be at all plausible they had better not imply that consciousness is something unique to humans.

Approaching the issue from a slightly different direction, if the idea is to see the soul as something evolved in the normal way (as Michael Ruse suggests in the book discussed in the last chapter), it cannot provide an immediate and sufficient basis for a radically different kind of capacity of the creature that possesses it. Or if it can, its acquisition at some particular point in evolutionary history, whether the point when ancestral hominids diverged from apelike ancestors, or the point at which some early creature dragged itself out of the slime, is something distinct from, and alien from, the normal course of evolution. It is not that evolution cannot endow an organism with a radically new capacity. This happens throughout the history of life. But evolution does so by gradual steps and continuous change.

Putting aside for the moment the sceptical conclusions of the last chapter, we should look for the most defensible conception of the soul in the Christian tradition. This, I think, is the tradition that sees the soul not as an independent substance but rather, in a way deriving from Aristotle, as something like the form of an animate creature (see Ruse, *Can a Darwinian be a Christian?*, 80 ff.). Aristotle makes a fundamental distinction between matter, of which things are made, and the form, which is something like the arrangement of the matter for a particular kind of thing, and which accounts for the characteristic properties of things of that kind. The relevant conception of the soul is developed by Aquinas:

The soul is the ultimate principle by which we conduct every one of life's activities, the soul is the ultimate motive factor behind nutrition, sensation and movement from place to place, and the same holds true of the act of understanding. So that this prime factor in intellectual activity, whether we call it mind or intellectual soul, is the formative principle of the body.

(Aquinas, *Summa Theologiae* 1a, 76, 1; quoted in Ruse, *Can a Darwinian be a Christian?*, 80–1)

On such an interpretation it is clear that all sentient creatures, perhaps all living creatures, will have souls, but not in the sense of having some detachable part, but rather in the sense of possessing a certain kind of complex organization and set of capacities. The divide here between humans and other animals will be one of degree, the exact extent of this divide depending, in turn, on differences that we shall consider shortly. Once we have accepted the idea of a God who takes a detailed interest in the activities of individual organisms on this particular planet, there is nothing additionally absurd about the notion that he might have a particular partiality for a group of creatures distinguished by a unique set of attributes. Perhaps as part of an ongoing overreaction to supernaturalistic ontologies, these differences are, I shall argue, often understated. So what I shall say from now on will tend to support doctrines of human exceptionalism.

As one last comment on religious views, however, it is worth noting that the conception of the soul just described is highly problematic for traditional conceptions of the survival of death. As the body decays it is hard to see where there might still reside any 'ultimate motive factor behind nutrition, sensation, and movement from place to place' since there is no longer any nutrition, sensation, or movement.

If the body is no longer animated there seems, as a matter of logic, no longer to be anything that animates it. In the end, it may well be that there are philosophical objections to the detachable, dualist soul that are as serious as those that face the Aristotelian/Thomistic soul as a vehicle for survival. The former do, at least, have the advantage of being less obvious.

The continuity of humans and other animals

There are several attributes that have been suggested as providing the most fundamental distinction of the human species. Language, thought, and culture come immediately to mind. There is nothing antagonistic to evolutionary thinking in the possibility of such special attributes. Many features of many kinds of organism are unique. The beaver is the only mammal that can digest wood, for example, and the platypus is the only poisonous mammal (the males have poisonous spurs on their hind legs). It is perhaps unusual for a wholly unique kind of feature to be restricted to a single species, but this reflects the lack of phylogenetic diversity of our lineage more than the special uniqueness of our species. If one considers more diverse lineages, unique features of the lineage are easy to find. Think of aerial echolocation in bats, for instance. Probably at one point in time there was just one species of echolocating bats. Imaginably, in the distant future there will be many species of talking, thinking mammals derived from our own species.

Let me turn, then, to the distinctively human feature that seems to me the most significant, language. I consider this more significant than either of the other two mentioned above for the following reasons. The only plausible ground for supposing that thought is unique to humans is the belief that language is. The kind of thinking emphasized by Gilbert Ryle, thinking that is revealed in action, is surely equally revealed by the intelligent tennis-playing human and the intelligent gazelle-stalking lioness. It seems improbable, on the other hand, that a lioness can engage in complex discursive thought for the obvious reason that it lacks the crucial instrument of discourse, language. Somewhat similarly, the most distinctively human parts of culture are surely those that depend on the human possession of language. There is a good deal of social organization in various other species, and those forms of organization that seem most clearly unachievable by non-humans seem unachievable in large part because they seem to require communication of a specificity that seems possible only with a complex language.

However, it should once again be emphasized that there need be no absolute discontinuity here. Humans are much more linguistically able than any other creatures we know, but so are blue whales much bigger. Whether or not it is also large enough to qualify as a difference of kind depends on the degree of difference. From this it follows that no conceivable threat to evolution is posed by insisting that the distinction between human and other languages is a substantial one. I emphasize the latter point because I do in fact believe that the uniqueness of human language is often underemphasized, sometimes because it is seen as a threat to the proper naturalization of humanity and its inclusion into the broader spectrum of terrestrial life. There is no such threat; and it is important, for reasons that I hope to make clear in the next chapter, that we take full account of the factors that are unique to *Homo sapiens.*

How large, then, is the discontinuity between humans and other animals with respect to linguistic aptitude? There is no question but that many other species use signs with what has been called 'non-natural' meaning. Such signs, that is to say, have no causal connection, or resemblance, to the objects to which they are used to refer. The connection between sign and thing signified, one is tempted to say, is conventional not natural. A well-known example is the group of signals used by Vervet monkeys to alert conspecifics to the presence of various different potential predators. There has also been a good deal of discussion of the degree of success achieved in experiments that aimed to teach symbolic languages to great apes and even whales. I don't propose to go into these debates here, however, for the simple reason that even the most optimistic assessment of the success of these experiments would leave a huge gulf between the achievements of our own species and that of any other.

To establish the evolutionary possibility of a trait unique to a particular species requires only that we can see plausible sequence of intervening steps between creatures possessing the fully evolved trait and creatures entirely lacking it. (Of course, we should not assume that our inability to conceive of such a sequence shows the impossibility that the trait evolved.) In the case of language there seems no particular difficulty here. As already noted, many animals have communication systems of varying degrees of complexity and sophistication. There is no obvious obstacle in principle to the massive development of such a simple communication system in the direction of human language. Of course a great deal remains mysterious as regards the forces that might have driven such an evolutionary trajectory, its relation to other developments in human cognitive complexity, and so on. But my point here is that there is nothing extraordinary about these problems, and surely no reason to imagine that we would need to go outside the normal framework of evolutionary thinking.

The principle deployed in the last paragraph, that we should avoid whenever possible postulating instantaneous leaps to states of entirely new kinds in our conceptualization of evolutionary trajectories, applies equally to our thinking about consciousness. Some philosophers—most famously Descartes—have held that consciousness was unique to humans among terrestrial life forms. This is not impossible to reconcile with a smooth evolutionary trajectory, as one may suppose that the first glimmerings of consciousness appeared at some stage of cognitive development beyond that of any other currently living creatures, but well before that which we have now reached. But I can see no possible reason for the counter-intuitive denial of consciousness to all other creatures. It seems obvious that we can tell whether a dog is conscious or not, and there is no need to question this simple bit of competence. A much better strategy would be to realize that consciousness is a quite complex concept admitting both of degrees and different domains of application. There are no doubt many things of which I can be conscious and of which my dog cannot—for example the fact that I owe a friend an apology for having forgotten her birthday. It is clear and important that such limits to the consciousness of a dog are intimately connected with its linguistic ineptitude. It is worth noting, though, that the dog, unlike me, may be conscious of the scent of a recently departed rabbit. The limits of a creature's consciousness are closely related to its particular set of capacities. As I shall elaborate in a moment, language provides us with an extraordinarily enhanced set of capacities, and consequently with an equally enhanced realm of consciousness. Perhaps this should be seen as a case in which a difference of degree amounts to a difference of kind. But if so, it is crucial to remember, from the point of view of evolution, that a difference in kind can be the summation of many small differences of degree.

The discontinuity of humans and other animals

The point I have stressed so far in this chapter is that there is no good reason to deny the evolutionary continuity of ourselves with other creatures. But, and this is the thesis that will occupy the remainder of the chapter, notwithstanding this continuity, the evolutionary novelty of language has had profound effects on even the biological status of the species to which it pertains. Most fundamentally these effects are due to the possibilities human language creates for the construction of elaborate cultures.

The significance of human language is at the same time so obvious and so diverse that it is difficult to discuss the point without banality. One approach to the topic that leads directly to the connection of language with culture is to note the vastly increased possibilities for division of labour, or division of role or status, that language makes possible for human societies. I take it as obvious that no non-human species even approaches this kind of differentiation. Two ways in which language facilitates this diversification are in the processes of training and in subsequent cooperation between agents with different roles. It is perhaps imaginable that the first function could be realized through some kind of internal developmental programme. Probably the most functionally diversified species other than our own are found among the social insects, and part of this is achieved through diversification of developmental physiology—though it is significant that signals from other members of the colony are crucial in determining the developmental trajectory that an individual will follow. At any rate it is perfectly obvious that in our own species it is through differences in training that one individual develops the skills, social role, and status of a baker, while another becomes a lawyer or a police officer. It is hard to imagine that these specific forms of training could occur without a complex system of transmission of expertise such as is provided by human language.

But even more interesting and fundamental is the need for communication between complementary roles within a complex system. It is useful here to compare a lower level of structural organization, the relations between the parts of a multicellular organism. It is still widely believed that the differentiation of cell types in the development of a multicellular organism is to be understood almost exclusively in terms of processes internal to the particular cell, directed by the genome. This view is coming under increasing and perhaps irresistible pressure. But what is beyond debate is that the successful functioning of such a system requires an enormous amount of communication between differentiated parts. The nervous system and the circulation of hormones and a variety of other biologically active molecules through the blood and lymphatic systems are the most obvious of such systems of communication. Communication between parts seems to be an essential prerequisite for the proper functioning of a complex system.

Returning to whole organisms, there is a great variety of communication systems that facilitate the interactions of different kinds of organisms. Any sexual organism is likely to have some form of communication between the sexes with which interests in sexual activity can be negotiated. Animals that live socially typically communicate in some way to negotiate conflicts or to establish status hierarchies within the social group. These are far from nonetheless hardly subtle enough to sustain the complexly interacting functions characteristic of human societies. Among the social insects cooperation and the division of labour is indeed sustained by quite sophisticated modes of communication. Most striking is perhaps the bee's waggle dance, the elaborate performance with which one bee informs its colleagues of the location of a valuable nectar source. But probably more significant in the end are the chemical communications between individuals that constantly direct all kinds of behaviour including the production of appropriate types of replacement individuals.

However impressive these systems may be, they do not approach the complexity and subtlety of human language, and the latter is surely a necessary condition of the complexity of culture that human societies enjoy and the diversity of roles that they exhibit. Certainly, to make a living as a soldier ant requires cooperative relations with other ants that provide one with food, and the smooth coordination of these activities requires some degree of communication. But to make a living in human society as a priest, a plumber, or a politician requires enormously more complex coordination in accordance with the very much greater number of roles. This line of argument is entirely neutral as to whether the coordination of social roles is to be understood as 'altruistic' in the sense that individuals actually care to promote the well-being of others, or whether it is entirely self-interested. Even if we were to adopt the most extreme economistic vision, in which no individual cares about the well-being of any other, and all interactions take place only if they are perceived by both parties as beneficial (or, of course, if they are coerced), there would still be the need for exchanges and mutually beneficial collaboration, and these transactions would require a complex medium of communication.

These are, needless to say, complex matters, and several of the issues touched on in the last few pages are the topics of numerous books. So it will be good to return to the reason why these matters are discussed here at all, and summarize this chapter. First, human language, like the giraffe's neck or the peacock's tail, has evolved to a state that may easily be seen as different in kind from the related features of any of its relatives. Nonetheless, there is nothing in this that should provide any trouble for the view that these features evolved naturalistically, by degrees, from some very different ancestral structure. But, second, as human language has evolved it has made possible other changes in human life that have even more profoundly distanced our own species from any of our relatives. Though I certainly don't accept that only humans are capable of thought, or even are conscious, there can be no doubt that the kinds of thought and the forms of consciousness of which we are capable are very different from those of other terrestrial creatures. And human culture, though not unprecedented, involves the articulation and synchroniza- tion of a variety of roles and functions that is different in kind from anything else in our experience. I am inclined to go further in elaborating these aspects of human unique- ness. For example I have argued elsewhere that the interplay between individual goals and social structures embedded in language provides a space in which can be found something that genuinely deserves to be called human freedom. I shall not attempt to reproduce this argument here. What is important for now is just to note that evolutionary continuity with the rest of life doesn't mean that there may not be features of human existence quite radically different from any found outside the human sphere.

6

Human Nature

Sociobiology

A main reason for the importance of the conclusion of the last chapter is that it casts doubt on the increasingly common idea that reflection on the process of evolution will profoundly illuminate human behaviour. This idea has appeared in a variety of guises ever since Darwin published his account of evolution, but its modern incarnation is generally dated from the publication by E. O. Wilson in 1975 of his monumental *Sociobiology: The New Synthesis.*

Wilson's book in fact contained twenty-seven chapters, of which only the last and a few provocative sentences in the introduction had any direct bearing on the human species. In the introduction Wilson predicted that evolutionary biology, developed in the ways he advocated, would explain (away?) ethics, not to mention 'ethical philosophers'. In the final chapter he sketched a range of insights into human nature that he thought could be gleaned from evolutionary reflections. The book, as perhaps intended by its publishers if not by Wilson himself, caused a furore. Wilson was accused of racism, sexism, and a good deal else besides, as well as bad science. A group calling itself the Sociobiology Study Group of Science for the People, and including some of Wilson's most distinguished Harvard colleagues, Steven Jay Gould and Richard Lewontin (whom, ironically, Wilson had gone to some trouble to bring to Harvard), published violent attacks on Wilson's work.

The heated debate over sociobiology left the topic in some general disrepute, though it also gained a growing band of dedicated followers. In the mid-1980s one group of followers began to organize itself around a version of socio-biology that they named evolutionary psychology, and which has become the most prominent contemporary version. Among the successes of this group has been to either ignore or caricature their critics, and to concentrate on estab-lishing the paraphernalia of a recognized and respectable scientific programme—specialist journals, conferences, postgraduate degrees, and the like. This appearance of scientific respectability, of what Thomas Kuhn in his classic text *The Structure of Scientific Revolutions* famously described as 'normal science', is illusory.

The causes of behaviour

Beyond a general sense that history is illuminating, why might one expect evolutionary reflections to cast light on human behaviour? A question immediately raised by this proposal is whether one thinks that behaviour is most usefully explained by considering features that the behaving agent brings to the table, or by considering the context in which the agent is placed. This may recall the age-old nature–nurture question but, for reasons that should become clear, I shall avoid using this terminology here, and refer more neutrally to questions of structure (of the agent) and of context. I shall assume, for the sake of argument, that the dispositions an agent has to behave in certain situations—for example the disposition to eat when confronted by a plate of oysters—can be understood as deriving from structural features of the agent, perhaps of the agent's brain. The context, on the other hand, will include such things as a dining room, a plate of tasty bivalves, a fork, and so on. This banal example suggests at once that explanation of behaviour cannot possibly get off the ground without appealing to both structure and context. I certainly cannot eat oysters unless there are oysters in my general vicinity. But the fact that I eat oysters when suitably presented with such, but refrain from eating mud if it is put on a plate in front of me or, for that matter, refrain from attempting to eat the plate, are dispositions that I bring to the table.

So much is straightforward. One might, however, minimize the importance of structure by suggesting that the dispositions that human agents possess are pretty much entirely determined by the context to which they have been exposed. Alternatively, one might imagine that these dispositions develop more or less independently of the particular experiences of individual humans. These possibilities bring us properly to the time-worn question of nature and nurture, and define more or less extreme positions within that debate.

Extreme positions are, indeed, a large part of the problem. In reflective moments almost everyone agrees that human dispositions develop as a result of interactions between the biological endowment of the organism and the environment in which it develops. As a simple and rather hackneyed example, there are biological facts that incline normal humans to learn a language, but which language they learn will depend on where they grow up. But despite these reflective agreements it is common for opponents to accuse evolutionary psychologists of biological determinism, sometimes suggesting the idea of behaviour as appearing quite independently of environment; and for evolutionary psychologists to accuse their critics of adhering to a 'blank slate' view of the human mind, a view according to which human dispositions are entirely unconstrained and un-affected by biology. To give its critics a more embattled and radical stance, this latter view has even come to be known by some evolutionary psychologists as the 'Standard Social Sience Model'.

The appeal to the Stone Age

It is, however, possible to discover some more serious dis-agreements underlying this exaggerated rhetoric. One such disagreement reaches fundamental issues in the under-standing of evolutionary processes. A standard argument deployed by evolutionary psychologists purports to show that the causal roots of human behaviour in the brain must inevitably be understood as adapted to the conditions of life in the Stone Age and must, therefore, be understood by reflection on the processes of evolution in the Stone Age. The superficially plausible argument for this position goes as follows. The brain is clearly an adaptive structure. It is a structure, that is to say, that has been fitted by evolution to serve the needs of the organism. But the brain is also a structure that, like any physiological structure, is built under the instructions of genes. Genes, therefore, must have been selected to produce this adaptive structure. But, it is argued, the selection of the genes necessary to produce a structure of this complexity will take a very long period of time, substantially longer than the time during which modern humans have existed. The longer period over which modern humans evolved from pre-human ancestors and, in particular, evolved their characteristically large brains, is generally identified as the Pleistocene, or late Stone Age. Hence, finally, it is this period of history to which the human brain is an adapted structure. Reflection on the conditions that human ancestors encountered in this period should provide the key to identifying the behavioural tendencies of contemporary humans.

The genocentric fallacy

One crucial premiss in this argument is the idea that adaptive features of organisms can only be permanently incorporated within a lineage if they are 'encoded' in genes. This remains something of a dogma of evolutionary thought, and is also associated with the gene-centred perspective on evolution made popular both among the general public and among certain streams of professional theory by the work of Richard Dawkins. It is, however, entirely mistaken and, indeed, perhaps the major obstacle to the progress of thought about evolution and a good deal else in biology.

The obvious and most widely acknowledged deficiency in this position is the possibility of cultural evolution. Patterns of behaviour may be imitated by conspecifics, and adaptive changes in a pattern of behaviour may be selected through the greater reproductive success of organisms that adopt them. Familiar examples of such extragenetically transmitted traits are birdsongs and, in many species, prey choices. In the human case this situation is pervasive. Humans learn a huge repertoire of behaviour from parents, teachers, peers, and other role models. Innovations in behaviour are sometimes directed specifically at adaptive ends and, whether or not this is so, they are often passed on to naive individuals. I make no claim here about the effectiveness of this kind of transmission in comparison to genetic transmission, but insist only that it is actual (and therefore possible). Somewhat ironically, in view of the way his work has been so widely understood, Dawkins explicitly acknowledged and discussed cultural evolution in his first and most influential book, *The Selfish Gene.*

It is important that non-genetic evolution is uncontroversially possible, and this is sufficiently established by the preceding point. However, the possibility of cultural evolution does not get to the heart of the difficulties with the gene-centred perspective that drives so much contemporary theorizing. The central problem is that the role of genes in evolution has been grossly misrepresented. Genes are still widely described as carrying blueprints for the organism, recipes for putting together organisms, and suchlike. Consumers of science fiction are often treated to the idea that with sufficient skill it would be possible to read off the features of an organism simply from a knowledge of the sequence of base pairs in the genome. At a more sophisticated level, many biologists still endorse the 'central dogma' that information about biological structure flows exclusively and unidirectionally from the genome. All of this can now be seen to be profoundly mistaken.

In the early days of genetics, genes were hypothetical entities identified as the causes of particular traits of organisms. When the structure of DNA was unravelled in 1953, together with the mechanism by which DNA replicated itself, it was natural to identify this earlier concept of a gene with stretches of DNA. When it was subsequently discovered that stretches of DNA directed the construction of proteins, it became common to think of a gene as the stretch of DNA that directed the production of, or 'coded for' a protein.

An immediate difficulty, however, was that the production of a protein is in general a great causal distance from the traits that had been the subject matter of classical genetics.

It is still very common to hear references to 'genes for' this or that trait—eye colour, intelligence, height, homosexuality, and so on. But it is vital to remember that though the production of particular proteins is necessary for the appearance of many traits, it is almost never close to sufficient. It has also become clear that most of the genome, the organism's DNA, does not code for any protein. Some parts are known to regulate the production of proteins, others have no known function and are sometimes referred to as 'junk' DNA. Finally, even those parts of the genome that do code for a protein do not typically specify a unique protein. Different parts of a gene specify subunits of proteins that can be assembled, sometimes using products of other genes, into a variety of proteins. Thus a better way of thinking of the genome is as a library of recipes. Which recipe is implemented, on the other hand, is often determined by features of the cell quite distinct from the nuclear DNA.

A good starting point for a proper appreciation of the function of genes, then, is the realization that the information needed to build an organism, far from being exhausted by the DNA, must minimally include an entire cell, which is in fact the smallest stage in the life cycle of any organism. The cell contains a great deal of material and structure without which DNA would be entirely inert and meaningless —its 'information' could not be 'read'. But the cell itself has a great deal of internal structure quite apart from that which 'reads' DNA. All of this is transmitted in reproduction and is essential to the development of the organism: there is much more to reproduction than the transmission of DNA. I have said that bits of the genome cannot in general be correlated with traits of the organism. This should be increasingly unsurprising as we come to realize the great variety of resources, even at the cellular level, that must be transmitted if a new organism is to develop.

Recent insights in molecular biology cut even more directly at the heart of the central dogma, at the idea that information travels only from, never to, the genome. It is now known that there are mechanisms by which the cell acts on the genome so as to affect the circumstances under which genes are expressed. The chemical basis of one important such mechanism is known as methylation. Technically, this refers to the addition of a methyl (CH_3) group to the C (cytosine) base where C is followed by a G (guanine). Methylation decreases the probability that a gene will be expressed. The best-known examples involve explanations of different probabilities for expression of genes that are inherited paternally and maternally as a consequence of different methylation patterns in male and female cells. There is speculation that such differences reflect evolutionary responses to conflicts of interest between mothers and fathers. But methylation also occurs throughout the life history of an organism and evidently plays a vital role in the processes through which different tissue types are differentiated in development. It is not always noticed that since methylation patterns are
_____ _____ ___ _____ _f genome sequencing this
provides one respect in which it is uncontroversial that the genome sequence inadequately specifies even the inherited biological resources of the organism.

These technical details are of great importance in under-mining what may seem profound reasons for overstating the centrality of the genome both to development and to evolution. Even more important for a general under-standing of biology, the conception of the genome as the sole repository of hereditary information about the organ-ism has served to maintain an ultimately disastrous rift between theories of evolution and theories of development. This conception has enabled evolutionists to 'black-box' development, as something somehow adequately specified by the state of the genome at any time. Evolution, therefore, could be seen as describing merely a sequence of genomes, without worrying about the messy processes that led from one genome in one generation to another in the next. The common theme to the most interesting work in contem-porary theoretical biology is the insistence that these must be brought together. One increasingly influential way of doing this is in terms of so-called developmental systems theory (DST), introduced in Chapter 2. DST insists that the smallest unit in terms of which evolutionary processes can properly be understood is the full developmental cycle from one stage in the life cycle through all the intervening stages needed to reproduce that stage in the next generation. On this picture, the genome is merely one developmental resource—no doubt a very important one—among others that are required in order to complete the various stages of the life cycle. Richard Dawkins's genocentrism gives a specific answer to the famous question about chickens and eggs: the egg came first. DST gives the more intuitively plausible answer, neither or both. The chicken is no more or less an egg's way of making another egg than is an egg a chicken's way of making another chicken. Bearing in mind the diversity of resources that must be found or reproduced in order to complete the life cycle, we may just as properly say that the bird is a nest's way of making another nest. Everything necessary for the reproduction of the developmental cycle is equally necessary for understanding the evolutionary trajectory of the organism.

Evolutionary psychology

After what may have seemed a rather lengthy digression, we are now in a position to see what is most fundamentally amiss with the central evolutionary psychological argument for seeing our species as atavistically adapted to life in the Stone Age, and hence as maladapted to life in our own era. It may be that the genome cannot have changed sufficiently for the transition from Stone Age life to contemporary urban existence, but the genome is only one among many of the resources that lead to the development of contemporary humans. Most obviously, the extra-organismic resources are very different, and in certain respects much richer. Schools, television, books, and so on did not exist a few centuries or millennia ago, and surely contribute to the development of contemporary human brains. No doubt the genome constrains the possible outcomes of human development. It may be impossible for us to acquire a social life of the kind enjoyed by ants or bees, for instance. But what these constraints may be is something to be discovered by empirical investigation of the diverse behaviour of different human groups, certainly not by speculation on what life may have been like in the distant past. On the whole, as anthropologists not corrupted by dubious biological theory have long insisted, the evidence is that there is a great deal of flexibility in human development.

So far in this chapter I have concentrated on rather general arguments for the necessity of understanding human behaviour in evolutionary terms. It is important to address such arguments because it seems to be very widely supposed that they show that something like evolutionary psychology must be the correct fundamental approach to human behaviour. It is, perhaps, an interesting hangover from theological cosmology that it should be assumed that attention to origins should be the key to understanding the nature of a thing. Certainly, if the origins of a thing can be traced to an intelligent designer the intentions of the designer are an excellent place to look for an understanding of the thing. And it is striking that the most ardent enthusiasts for evolution as a source of contemporary wisdom do indeed go to great lengths to reintroduce the concept of design. But of course design is no more than a metaphor in application to organisms, and it appears to be an extremely dangerous one (it might perhaps be named Dennett's Dangerous Metaphor).

The (lack of) evidence for evolutionary psychology

A further reason for addressing very general arguments for evolutionary psychology is that it appears that only the conviction derived from these general considerations can explain the derisory quality of evidence that seems acceptable in this area for more specific claims. It is important to emphasize the difficulty of the task evolutionary psychology sets for itself. When an evolutionist sets out to explain the length of the giraffe's neck there are serious pitfalls to be faced, as I have discussed in earlier chapters. But at least it is not open to serious doubt that giraffes have long necks. By contrast, the evolutionary psychologist is typically advancing a thesis about human nature at the same time as offering an explanation of the trait hypothesized. So, for example, on the basis of quite abstract arguments about the subversion of cooperative arrangements by defectors, and the evolutionary importance of detecting such subversive attempts, it is proposed that the human mind is specially adapted to detect attempts to defect from social rules. In a terminology favoured by these theorists, it is proposed that there is a special module for detecting cheats. Unfortunately, as in most such cases, it is perfectly evident at the outset that people have some interest in detecting violations of social rules, and good and obvious reasons for such interest. It will not be an easy task to provide a convincing g behaviour conducive to this interest should be explained not merely as a clearly sensible way to act, but as a kind of action driven by a specially designed part of the mind. If, on the other hand, one has convinced oneself on a priori grounds that something of the sort must be true, the project is likely to seem a great deal easier.

There are some more or less evidence-based strategies of argument offered for the proposition that contemporary human behaviour reveals its origins in Stone Age—or other long distant—conditions, and I shall consider some of these in the concluding parts of this chapter. One common strategy in popular presentations of evolutionary arguments is animal comparison. Male red deer or elephant seals fight one another to the point of fatal injury in competition for access to more females, so perhaps men are disposed to do the same. Female birds of many species are observed to sneak off for sexual liaisons with males other than their primary partners, so probably women are similarly inclined. Males of various species are observed to kill the children of female partners with other males, so we should be unsurprised that human step fathers display violence to their non-genetic children. And so on.

The irrelevance of these comparisons may seem obvious enough from the simple observation that the comparator species need generally to be carefully chosen: many other species display no relevantly similar behaviour. However, the argument might be interpreted differently. Perhaps the comparisons show that the behaviour in question is the sort of thing that can evolve through natural selection, and so if we find some instances of the behaviour in question in humans we have good reason to suppose that a relevant tendency has evolved. A first response to this version is to note an absolutely standard point in evolutionary thinking, the distinction between analogy and homology. Homology refers to the situation in which different species have a similar trait through descent from a common ancestor with an ancestral version of the trait. So, classically, the flipper of the whale and the wing of the bat have similar bone structures, and it is believed that the reason for this is common descent from a very different but skeletally similar ancestor. The traits are homologous. On the other hand, the wing of a bird and the wing of a bat may be very similar in some respects, reflecting the similar selective pressures exerted by the laws of aerodynamics, but these similarities are mere analogues, as they have certainly evolved independently from one another. Tracing back from either species to a common ancestor one must pass through many species that have none of the relevant traits, so we can be sure the traits evolved independently.

Almost without exception, the parallels offered in support of human evolutionary psychology are, at best, examples of analogy. And pointing to analogies can tell us nothing about the actual evolutionary trajectory of a trait. They can, no doubt, tell us that there is some evolutionary tendency to acquire such traits, but also, given the contrasting species just noted, that this tendency is contingent on other factors. But, of course, to the extent that the traits in question are at least part of the repertoire of human behaviour, there is no doubt that they appear under certain circumstances. If we use the term 'evolution' in the broadest possible sense, then we can say that they have evolved. The problem is, as I have argued in some detail earlier, that the broad sense of evolution includes the kind of cultural processes that it is precisely the aim of evolutionary psychology to reject. How can we decide between different kinds of evolutionary process as explanations of particular behavioural traits?

The proper realm of genetic explanation

Bearing in mind that the core of the general argument for evolutionary psychology puts genetics at the centre of the explanation of human behaviour, we should ask what distinguishes phenomena that are appropriately explained in this way. A natural thought is that we should explain phenomena genetically if they develop in ways that are largely insensitive to environmental contingencies. Starting with the idea that features of organisms develop through a process of interaction between biological and environmental factors, we might think to emphasize one or other kind of factor to the extent that the phenomenon in view is largely independent of the impact of the other. Whether a child learns to speak French is largely independent of their genetic peculiarities and almost entirely determined by the environmental fact that they grow up in a French-speaking country. That they have blue eyes, on the other hand, seems to have little to do with environment and to follow reliably from having a certain genetic inheritance. One should note, however, that very many features of organisms cannot be classified in this way. Consider, for instance, the fact that the vast majority of humans learn to speak some language or other. This tendency may be derailed by sufficiently abnormal biology, for example genetic abnormality, but also by abnormal environmental conditions, notably the lack of an ambient language in the child's environment. Biologically normal children in normal human environments will learn a language, but it seems wrong to attribute this fact either to biology or to environment. Perhaps less obviously, the same thing should be said of such typical developmental outcomes as having four limbs. If all goes well this is how people come out but, as with language, either unfortunate biology or mishaps of environment can prevent this outcome. In yet other cases, such as the development of a functioning liver, we have features without which the organism will not develop at all. But even here properly functioning genes as well as environmental factors such as adequate nutrition and shelter are equally required for the successful development of these traits.

What appears to follow from this is that genetic explanation is appropriate for features that appear in some but not all members of a species, and for which the difference between its appearance and non-appearance is attributable to differences in genes. This is indeed the object of investigation in the sciences of behavioural genetics and, much more abstractly, population genetics. Evolutionary psychology, however, follows a rather different strategy. Most often it seeks to identify typical or universal features of human psychology and then to claim that these are to be explained genetically. Typical or universal features of the members of a species should in fact be explained developmentally. It is of course true that most features will require normally functioning genes to develop properly. But, as emphasized above, they will require a great deal else besides. What is of interest short of a full understanding of development, a distant goal, is to know what variable aspects of development might affect or even prevent the appearance of the feature. Evolutionary psychology often conveys the impression that only genetic variables are likely to affect or prevent typical psychological features of humans. But they have no cogent argument for this claim. Their allegations are thus not only misleading but also, because they tend to discount the possibility that psychological features might be largely dependent on environmental factors and therefore susceptible to change, harmful.

One might misunderstand the nature of evolutionary psychology's claims by focusing on the fact that the features they claim to explain are often quite uncommon ones. So, for example, one notorious recent thesis has been the evolutionary explanation of rape. Since most men do not rape one might imagine that this was indeed the explanation of a variable feature in terms of variable genetics. But this would be a mistake. Evolutionary psychology does not suggest that some men have mutant genes that cause them to rape, but rather that it is typical or normal for men to have a disposition to rape. This is said to be part of a variable set of sexual strategies. That most men don't rape is a consequence of the fact that most men are fortunate enough not to find themselves in the circumstances (low status and hence little access to willing females) in which rape becomes the optimal strategy and, in that context, is biologically determined. My own view (though I must admit that there are radical feminists as well as evolutionary psychologists who would differ) is that most men have no disposition to rape. Indeed, they may have no disposition violently to coerce anyone into doing anything. The disposition is a defective developmental outcome and, though it is possible that it is the result of faulty genes, it is an overwhelmingly more plausible hypothesis that it is more commonly a result of substandard environments. More to the point, rape as a genetic abnormality would not work for the theoretical purposes of evolutionary psychology. Being an obligate rapist is very unlikely to be an evolutionarily successful strategy, and no evolutionary psychologist of whom I am aware has suggested that it could be one. If the disposition to rape is a developmental failure, whether genetically or environmentally caused, there should be no selective explanation for it and it is outside the scope of evolutionary psychology.

The preceding example, though admittedly on the fringes of respectability even for evolutionary psychology, points to many of the main deficiencies in that pseudo-scientific project. It is based on a simplistic genetics informing a naive conception of the evolutionary process. It entirely overlooks the processes of development, the black box inside which it would be sensible to look for the sources of good and bad developmental outcomes. And by giving a spurious causal explanation of many undesirable developmental outcomes (violence, greed, sexual predation, and so on) in a realm beyond the influence of either social or individual control, it is pernicious in discouraging serious investigation of how more satisfactory developmental outcomes could be obtained.

Can knowledge of evolution then tell us nothing about what we are like? My suggestion is that at the level of specificity sought by a project such as evolutionary psychology, the answer is almost certainly not. A simple way of reinforcing this conclusion is merely to reflect on the diversity of the products of evolution. All have evolved, yet the results are almost inconceivably diverse. So the fact of evolution can tell us little about its products. It is true that there are important signs of the common origins of different terrestrial life forms. Ironically, the most striking of these is the genome. Not only do the most diverse organisms share the same basic structure of the gene and pattern of translation from DNA to protein (the genetic code), but it turns out that the complement of genes themselves is surprisingly similar. It is often remarked that our genomes have turned out to be 98.4 per cent identical to those of chimpanzees. We are invited to conclude that we are, contrary to our inflated expectations, 98.4 per cent identical to chimpanzees. But if this means anything (which I rather doubt) it is surely false. The correct inference is, of course, that neither we nor chimpanzees are identical to our genomes. That this conclusion is not usually drawn speaks volumes for the contemporary power of gene mythology. More generally, and this is the irony referred to a moment ago, to the extent that genomes are among the most invariant of features of different organisms, they are the last place we should expect to find explanations of the most specific features of organisms. The details of the human mind, in particular, are the parameters of a feature largely unique to a particular species.

The crucial point that the diversity of evolutionary outcomes mandates is empiricism. Perhaps the deepest fault with evolutionary psychology is its attempt to infer human nature from theoretical principles rather than describing it on the basis of observation. Occasionally it is said that evolutionary psychology is no more than a device for generating interesting hypotheses for investigation, and in that role it is largely harmless, though my own reading of the record doesn't suggest that it has been very fruitful. But this modest statement of ambition is disingenuous. This is clear from the paucity of evidence that is often taken to confirm the hypotheses put forward by evolutionary psychologists. It is not that there is no such evidence. Indeed, the hypotheses are often so banal that it would be extraordinary if there were none. That men often tend to polygamy and occasionally commit rape; that people are sensitive to and often object to violations of social rules; or that step parents often treat step children less well than their biological children, to take some of the more central claims of evolutionary psychology, are hardly news. That these generalizations are confirmed by some research does not tell us anything at all about why they are, to the extent that they are, true. Only a conviction that the methods touted by evolutionary psychology embodied some profound wisdom could lead anyone to imagine that these investigations provided the path for a proper understanding of that most frustrating object of scientific enquiry, the human mind.

7

Race and Gender

The previous chapter addressed the relevance of evolutionary thought to particular kinds of behaviour. A different, if related, issue is the relevance of evolutionary ideas to the division of people into kinds. The last chapter also considered the division of labour that is characteristic of all human societies to some degree. While this is clearly a good thing in so far as it allows great gains in efficiency (as Adam Smith demonstrated with his iconic example of the pin factory), it is also the basis for all kinds of less positive associations of people into kinds with differing status. Whether or not this constitutes an avoidable evil, something that surely does is the association of categories of work with categories of people defined biologically or quasi-biologically. Glass ceilings for women and limited opportunities for people of colour are familiar and pervasive problems in Western societies. Defenders of these prima facie inequities often appeal to biological differences between groups as a justification for them. While it is important that we do not see the legitimacy of discrimination as contingent on the outcome of biological debate, it is of some use to point out that the current state of biological knowledge provides little or no reason to think that groups distinguished by sex or race have systematically different capacities that might explain their different places in the division of labour. This chapter will consider what reflection on evolution can tell us about these categories and, more importantly, what it cannot. The biological status of the two kinds of category with which the discussion will be concerned is very different, and they will be considered in turn. From a biological perspective race can be dealt with more easily. Sex (or gender: the difference will be considered below) is rather more tricky.

Race

It is universally agreed that there is only one human species, but for a long time it was supposed that there were biologically significant subgroups, probably not with the full status of subspecies, but rather what are known within biology generally as geographical races. On the other hand, it is common nowadays for biologists to assert, on the basis of genetic investigation, that the concept of human race has been shown to be meaningless. Richard Lewontin and Stephen Jay Gould have been especially prominent in emphasizing that humans are a relatively homogeneous species genetically, and that the variation there is exists within much more than between biological populations. There is much more genetic variation within a racially defined group than there is between any two such groups. On the assumption that the purpose of classifying a thing is to convey information about that thing, this observation suggests that classifying by race is, at least from a biological point of view, pointless or even meaningless.

Matters are a bit more complicated than this, however. To make any sense in this area we need to begin by distinguishing race as a biological concept from race as a sociological concept. There is no doubt that the latter is of great importance in many societies. Status, access to social goods (most obviously desirable jobs), and many other things that matter to people are demonstrably affected by the racial group to which society in general assigns people. Typically this assignment is something that members of societies learn to do largely unthinkingly on the basis of a few superficial characteristics. Where racial classifications are incorporated into official regulations, there are often complex criteria used in assigning problematic individuals, for example individuals with various racial mixes of ancestry. Given this idea of sociological race, we can then ask the question whether there is any significant biological basis for this concept. And the answer is decisively negative. There are, no doubt, typical genetic correlates of the features that are used to make racial judgements—mainly skin colour and facial features— and that is it. Skin colour, almost certainly an adaptation to cool or sunny climates, is a superficial characteristic, quickly evolved and lost as groups of people move from one climate to another, and is a characteristic that has been evolved by many distinct groups at different times and places.

However, we may still wonder whether there is any other

concept of race that is of more serious biological interest. An interesting argument by the biologist Massimo Pigliucci and the philosopher Jonathan Kaplan suggests that there is. Their idea can be approached through consideration of the finer texture of human evolution. Since there is only one human species, any splinters from the human lineage that reached full separation from the parent stock have evidently become extinct. Nevertheless, within a species as large and geographically extensive as *Homo sapiens* we can be sure that there have been, at different times, temporarily more or less isolated populations that will have acquired some degree of specific adaptation to their particular ecological circumstances. As a matter of fact recent work on the evolution of resistance in plants to local toxins suggests that isolation from surrounding populations is not required for such local adaptation if local selection pressures are strong. These variations, well below the level of distinct species, are referred to as ecotypes and are familiar in many different species. No doubt in some cases a distinct ecotype will be the first stage in the progression towards distinct specieshood. But in a species such as our own, given to migration and exogamy, this seems not to happen. What we do have, according to Pigliucci and Kaplan, are many more or less distinct ecotypes or their remnants scattered through large polytypic populations.

The first thing to note about this account is that it has little or nothing to do with the sociological concept of race. The main reason for this is that it is very much more fine-grained. Most sociological concepts of 'black', for instance, will include not only quite diverse peoples of African descent but even indigenous Australians. These will certainly form a large number of distinct ecotypes, and many of these ecotypes will be biologically closer to some ecotypes not classified as black than to other black ecotypes.

Given that this is such a different concept from the sociological concept of race, it may well be wondered what is to be gained by promoting it. One answer, with which I have some sympathy, is simply that it is best to get things right. But this account also helps us to understand better some puzzling aspects of racial difference. For example, whereas it is often said that there is no biological basis to race, given the intuitive obviousness of racial categories to many people, this assertion often fails to carry conviction. Pigliucci and Kaplan's more complex account helps to explain the phenomenon that strikes people as obvious while at the same time making it clear that the intuitive concept of race brings together diverse groups of people that may have nothing more in common than their membership of the human species. Without supposing for a moment that the presence of such an account will immediately dispose of all or even any of the deep social problems associated with racial classification and discrimination, it seems to me a small step in the right direction.

There is also a more subtle reason why it may help to defuse some of the intuitive grounds people have for taking broad sociological racial categories too seriously. Against the claim that biological investigation shows nothing much of interest to be common to all people classified as black, the casual observer is likely to observe, for example, that black people are demonstrably more successful at some sports, and this presumably goes deeper than skin colour. If blacks are more athletic, why shouldn't white people be better at thinking, as well-publicized claims about race and IQ are sometimes taken to suggest? As Pigliucci and Kaplan point out, the problem with 'blacks are more athletic' is that the real information is swamped by the large and meaningless classification. It is probably true that people from a certain part of Kenya are a good deal better at running marathons than most other people. An ecotype of Kenyan descent may well have developed extraordinary physical stamina. And there are very probably other black (and no doubt non-black) ecotypes with a tendency to particular physical aptitudes. The mistake is to interpret these very local biological categories in terms of large and amorphous racial categories.

It should be said that there are further issues in the debates over race and IQ. First, a great deal of the more notorious work in this area is deeply flawed, especially by misapplication of the very tricky notion of heritability. This is a pervasive problem in the popular assimilation of genetic information. We are constantly exposed to such claims as that research has shown that intelligence is 60 per cent genetically determined. Usually the basis for such a claim is an estimate of genetic heritability. Roughly speaking, what genetic heritability measures is the proportion of variation in a continuously variable characteristic (height, score on an IQ test, longevity, etc.) that can be explained by differences in genes. Something to notice at once is that if we were to enforce a totally homogeneous environment, perhaps through some extreme totalitarian government, all variation would have to be explained by genetic differences — the only differences there would be—and the genetic heritability of all traits would be 100 per cent. Conversely, in a population of genetically identical clones, the genetic heritability would be zero. It should be obvious that genetic determination is not at all what heritability measures: if intelligence is genetically determined, presumably it would be determined even among a population of clones. In fact, as far as I know, the expression '60 per cent genetically determined' is entirely innocent of any meaning whatever. A high level of heritability says nothing about the degree to which the transmission of a trait is biological.

The preceding point emphasizes the fact that it is extremely difficult, very likely impossible, to separate the biological and social causes of a complex variable such as intellectual attainment. As I explained in the previous chapter, the development of complex characteristics is a result of a continuous interaction between biological endowment and environment. In any society in which race is a significant factor—all or most current societies, that is—the environment will systematically be different for people of different races. Growing up as a member of a disadvantaged group is bound to be a different experience from growing up as a member of a dominant group, and the differences are likely to be so pervasive as to make it close to impossible to control for them in any meaningful investigation of differences abstracted from race. So sorting out a biological component of this difference will be somewhere between very difficult and impossible. To some degree the same may be said of athletic attainment. Differences in aptitude here, however, may be both more clear-cut and more directly related to features that were subject to strong selection among human ecotypes, suggesting that differences grounded in the particularities of distinct ecotypes may have some significance.

Before leaving the subject of race, it is vital to stress that none of this has any bearing on the justification for treating individuals differently on the basis of race, for there is no such justification. Differences in average IQ, if these were established, would be entirely irrelevant to the decision whether any individual was suited to a particular position. If IQ is relevant to the suitability of a person for a job, say, it is the IQ of the individual not the average IQ of a group that matters. There is some risk that a discussion of the kind just presented will be taken as implying that the justification of group-based discrimination is an empirical issue. This would certainly be a disastrous misunderstanding: the demands of natural justice are in no way hostage to empirical findings. But I hope that the risk of such misinterpretation is worth taking in the interest of trying to be clearer about the biological reality in question.

Sex

By contrast with the subtle and controversial biological basis of race, there is no doubt whatever that there is a solid biological basis to sex. It is true and very interesting that we are inclined to insist on a much more rigid and absolute difference than biology really provides. A significant number of human infants are born with features characteristic of both sexes, and the appearance of exhaustive sexual categories is partly an artefact of the insistence by society and the medical profession that every infant be assigned to one sex or other, even if this requires extensive surgical intervention. Still, a fluid boundary between the sexes doesn't contradict the fact that most cases are unproblematic, any more than the fuzzy boundary between the bald and the hirsute provides a bald man with a full head of hair.

It is an interesting fact that the evolution of sex remains poorly understood. If one thinks of sexual reproduction as evolving within a previously asexual species, sexuality looks to be evolutionarily a very bad idea. This is one place where the issue can most easily be seen from the gene's eye view. An organism in a population of asexual reproducers that attempted sexual reproduction would only pass on half its genes to each of its offspring. Unless it was able to produce twice as many offspring, its genes would do badly in evolutionary competition with its asexual rivals. Consequently one would expect the genes tending to produce sexuality to die out rapidly. One could imagine a species in which collaborative males and females were able together to produce more than twice as many offspring as single parthenogenetic females. However, the reality is that in most species there are sexually reproducing females raising offspring pretty much unaided. Why are they not out-competed by parthenogenetic mutants who can pass on twice as many genes? Put differently, from an evolutionary perspective males who contribute little to rearing offspring look like genetic parasites. Why doesn't evolution do a better job of getting rid of them?

The answer must presumably involve some very considerable benefit of sexual reproduction. It used to be popular to suggest that this genetic recombination between partners would provide variety and thus enable a species to respond more efficiently to environmental changes or opportunities. But this is now seen as a very shaky argument. Most genetic changes are harmful, and from the point of view of an individual with a well-adjusted combination of genes, good enough to get one to the point of trying to reproduce, the last thing one is likely to want to do is shuffle one's genes in the hope of something even better. It is possible to argue that there are selection processes at the level of species, and those species better able to evolve will in the end survive better than those stuck with a more inflexible genetic endowment. But selection between species is a controversial idea, and it is easy to see why. It's all very well having the possibility of evolving into some splendid new type, but if the species has already died out in a flurry of unsuccessful genetic experiments it will probably never happen. It appears that sexual reproduction must have some rather more immediate advantage.

The most plausible candidate for such an advantage is the ability to resist the predations of micro-organisms. Microscopic parasites, with generation times of minutes rather than years, can evolve thousands of times faster than their multicellular hosts, and within the lifespan of the host can evolve to exploit the environment provided by the host with ever greater efficiency. If offspring were genetically, and hence biochemically, identical to their parent, they would be born equipped with perfectly adapted parasites waiting to prey on them. Sexual reproduction, by providing subtle differences in the biochemical environment, gives the organism a new jump on the parasites each generation. Or so the story goes.

At any rate, we know that sexual reproduction is a given within the human species. Probably a few humans will be cloned some time soon, perhaps by the time this book is in print. But even if, as is far from certain, this technology proves capable of producing healthy humans, the expense and difficulty will surely leave sexual reproduction as the overwhelmingly dominant mode of reproduction for the foreseeable future. Whatever the selective advantages of sexual reproduction may be, sexual organisms evolved at some point from asexual ancestors and, judging by the frequency with which it is now observed, sex has proved to be a good idea.

Sex and gender

The question that will occupy the remainder of this chapter is whether the evolutionary origins of sex can tell us much about contemporary differences between men and women.

It will be useful to begin by referring to a distinction developed a few decades ago by second-wave feminists, but sadly falling increasingly into misuse, between sex and gender. 'Sex', as I shall use the term, refers to a biological difference between male and female. Paradigmatic male humans have penises, facial hair, etc., and an XY chromosome pair; paradigmatic human females have vaginas, uteri, and an XX chromosome pair. Some people, as noted, do not fall readily into either of these sexual categories, but a large majority do.

'Gender', on the other hand, refers to something quite different, the systematic differences in behaviour between male and female humans. Unlike sex, gender appears to differ radically across human cultures. In more traditional societies gender is quite rigid. Men and women are educated differently; are expected to end up in different occupations, women often being expected to devote themselves to domestic work and child-rearing; wear different clothes; spend time in different places; and so on. In contemporary Western societies gender differentiation has become a good deal more fluid. There are, however, strong statistical differences. Boys and girls tend to play with different toys, and boys play more roughly. Many occupations remain dominated either by men or by women. Much more domestic work is still done by women, and women spend a lot more time caring for children as well as the sick and the old. Although there is increasing fluidity, clothes are still fairly strongly differentiated. Women wear a large majority of the lipstick and high heels, men a majority of the ties and boiler suits. And behaviour is judged differently according to the gender of the agent. Aggressive behaviour that is perceived as indicative of drive, vision, and ambition in a man may be judged much more negatively in a woman. A great deal of research, mostly inspired by feminism, has explored the ramifications of gender division across all facets of society. It should perhaps be mentioned that feminist scholars have not been uncritical of the distinction between sex and gender, and some have argued, for instance, that it falsely suggests that sex is beyond the reach of any kind of social influence. Without dismissing these doubts, for present purposes it would be progress enough if the distinction were properly understood.

It is important to note that sex and gender can come apart. Aspiring transsexuals are typically required to adopt their desired gender for a period of time before undergoing the surgical and hormonal treatments which, on some views, will bring about a change in their sex. (One of the ways in which the terms 'sex' and 'gender' are sometimes now most strikingly misused is in talk of 'gender reassignment surgery'. One attempts to reassign one's gender by changing one's clothes and behaviour. Surgery may or may not succeed in changing one's sex.) History records cases of human females who have successfully pursued careers in professions open only to men, and have done so by adopting successful and largely unquestioned male gender. Still, these are exceptions, and men and women divide according to gender along the lines of the biological distinction of sex.

Gender and evolutionary psychology

Because of the cultural differences in the articulation of gender, and because of the increasingly fluid gender divisions that have become prevalent in modern societies, not to mention the cases mentioned in the preceding paragraph, it has become common to assume that gender is fairly loosely connected to sex, and is largely a social imposition on the basic biological division. However sociobiologists and more recently evolutionary psychologists have strongly objected to this tendency. And it is easy to see why they should do so. From an evolutionary perspective the most important thing an organism can do is to reproduce. Natural selection is often described in terms of an imperative of survival and reproduction, but it is always understood that the former is only a means to the latter. There is no evolutionary advantage in living as long as Methuselah if one produces no offspring. Since males and females play different parts in reproduction and must pursue different behavioural strategies if they are to reproduce, it is natural to assume that sex will be correlated with different behavioural dispositions. This is demonstrably the case for most sexual organisms, and so extrapolating on the general principle of treating humans as broadly continuous with other organisms it is natural to assume the same for us. It is thus natural, finally, to assume that gender differences are basically an articulation of biologically mandated strategies for reproductive success.

The first step in articulating the foregoing perspective is

to reject the impression that gender is highly variable across cultures and argue that really gender differences are slight variations on a universally mandated general human pattern. What pattern? There is a laborious approach, which would be to attempt to analyse numerous cultures, study their gender differences, and discover underlying communalities. This is, however, an unpromising strategy. There are countless different things people do and any of these may be done in subtly different ways by men and women. Do we focus on the way people dress or the way they chew their food, for instance? Much more promising is to work out through general evolutionary argument what commonalities to expect, and then see whether cultures can be found to show only minor variations on those themes.

Here we find a classic argument that occurs in every standard treatment of sociobiology and evolutionary psychology. The argument analyses reproduction from an economic perspective, in terms of returns on parental investment. The starting point is the observation that the minimal investment is much less for a man than for a woman. Men can, with good luck, achieve reproductive success in a few seconds, whereas minimally women must spend nine months pregnant and usually many more caring for an infant or child. Having invested their few seconds, men can choose between helping with the childcare and looking for further reproductive opportunities. Given this lack of further commitment, men will be willing to have sex at every opportunity, while women, it is argued, will only be interested in outstandingly good opportunities, in practice either outstanding genes or a credible willingness to help look after the kids. The optimally evolved woman will aim for both: get a sucker who thinks you're bearing his children to take care of them, but sneak off in search of the best genes when he's not looking (or perhaps tied up with the children). In support of this Machiavellian account of female psychology it is claimed that female infidelity usually occurs at the most fertile point in the menstrual cycle, and it is alleged that perhaps 15 per cent of human children do not have the genetic father they are thought to have. The last observation invites disturbing speculations about the likely consequences of introducing cheap and reliable genetic testing for paternity.

This basic contrast is elaborated into rich theories of human behaviour, and large research projects attempt to demonstrate the universality of behaviour adequately fitting the alleged pattern across as many different cultures as possible. Not surprisingly, the behaviour proposed fits many Western stereotypes. Men are aggressive, promiscuous, risk-takers, and are attracted to young healthy women with hour-glass figures—indicative of lack of previous children and, apparently, good health. Women are cautious and sexually manipulative—offering sex only in exchange for childcare or food, unless the man is genetically irresistible. They are attracted to men for their power and resources. And so on.

I won't go into much detail in this chapter as to why I find this kind of theorizing wholly unconvincing, because I have said so in general terms in the preceding chapter, and what I said there applies fully to this central case. Here I shall put the main argument of that chapter in a slightly different way. Because they are stereotypes, many people will find the accounts of gender offered by evolutionary psychology convincing. Stereotypes often have a basis of truth. Although there is great diversity in sexually differentiated behaviour in the animal kingdom generally, there are also patterns, and many species fit the stereotypes offered for humans. Ambitious male sociobiologists often appear to find the image of dominant male elephant seals or stags holding their rivals at bay and enjoying access to harems of females intuitively relevant to the human condition. All I want to argue here is that even if human gender roles were an evolved correlate of human sex, and a correlate evolved by some part of the human lineage ancestral to all contemporary human populations, this would tell us nothing at all about the mutability of gender roles. The importance of this conclusion is, I hope, obvious enough. Biological explanations of social facts are frequently interpreted as having conservative implications. If it's part of our biology, the thought goes, we might as well just learn to live with it. No such implication is necessary, however.

Gender as part of culture

The basis for my counter-argument will _____ _____ surprise. The conservative reading of evolutionary claims comes about from the assumption that evolution involves essentially the accumulation of genes. But on the view advocated in this book this is by no means the only way that evolutionary change can occur. And, once this is noted, it is a most unlikely route for the evolutionary processes currently under consideration. We should take seriously the apparent diversity of sexual behaviour even if we suppose there is a substantial underlying genetic influence on it. We need only assume, in accord with what is correct about the central evolutionary psychological argument, that the recent divergence between human populations is driven by changes in the cultural rather than biological components of the aetiology of gender difference.

No one questions that human development, just as the development of any other organism, will normally result in a strong disposition to engage in sexual activity. There is almost equally little doubt that the kind and frequency of sexual activity engaged in is strongly dependent on social context. Victorian country gentry almost surely engaged in less sexual activity than, say, contemporary British holiday-makers in Ibiza, and not because of any difference in their genes. Given then that behaviour is dependent on culture, we can ask whether either past, historically recorded changes in behaviour, or possible future changes, are likely to be the result of cultural changes or of genetic changes. Again the answer is obvious: human culture is much more rapidly mutable and flexibly responsive to its situation than is the human genome.

One might agree that human sexual behaviour changes with cultural changes and varies substantially across different cultures, and still maintain that there is a constant difference between male and female sexual dispositions. For example, it might be suggested that males are consistently more inclined to promiscuity. An obvious difficulty with this hypothesis is that sexual activity requires two participants and the predominance of male over female homosexuality is not great enough to threaten seriously the conclusion that males and females engage in about equal amounts of copulation. It might be that women are less enthusiastic, though this would seem to me a difficult hypothesis to establish. No doubt the prevalence of prostitution indicates that some female sex is motivated other than by sexual desire, and no doubt there are circumstances short of literal prostitution in which women engage in sex for a variety of instrumental reasons. It is also possible, and perhaps plausible, that many men would like to engage in more sexual activity than they in fact do. So we may allow the possibility that there is some systematic difference in disposition to sexual activity.

Still, this gives us very little or no basis for predicting how people will behave in any actual situation. It is evident that both men and women are strongly inclined to have sex sometimes, and in appropriate circumstances both will have quite a lot of it. We must also bear in mind how difficult it is to distinguish the biological from the cultural bases for whatever differences there may be in a disposition and here, as usual for fairly obvious reasons. In all cultures women will be aware that sex may have very serious consequences (pregnancy and responsibility for an infant), and thus a higher threshold of willingness to take the risk is to be expected quite apart from any biological differences. As one curious final observation, given the role that the standard insistence on our continuity with other animals reliably plays in these discussions, it is ironic that among one of our two closest relatives, the bonobos (or 'pygmy' chimpanzees), males, females, young, and old engage frequently in sexual relations of all kinds (homosexual, heterosexual, genital, oral, manual, etc.). It has been suggested that bonobos are unique in the animal kingdom in the extent to which casual sex is a fundamental part of their social interaction. Perhaps the human species also shares some of this divergence from the traditional predictions of sociobiology.

In summary, then, we know that what determines actual human sexual behaviour is largely cultural; the histories through which diverse human sexual practices diverge should certainly be understood as cultural rather than biological evolutions; we have no idea how to partition what systematic differences there are between male and female behaviour into biological and cultural components. Biology, and hence more specifically evolution, is of very limited use to us in understanding sexual difference. Possible articulations of gender difference are of course constrained, as everything we do is, by our biology. But evolutionary reflection is unhelpful, and often dangerously misleading, as a resource for understanding this phenomenon.

8

Conclusion

arlier in this book I quoted Theodosius Dobzhansky's famous remark that nothing in biology makes sense apart from the theory of evolution. I don't intend that anything I have said contradicts that famous dictum. However, and hardly surprisingly, much in biology, and more especially human biology, needs a good deal more than evolution if it is to make sense.

It is a thesis of this work that the importance of the theory of evolution is often misunderstood. Many scientists, rightly impressed by one of the most significant advances ever made in our understanding of the world we live in, try to get more out of Darwin's theory than it can provide. Evolution can't give us detailed explanations for the countless features of organisms. One important reason for this is that these features are truly countless: there is no limit to the number of features we can distinguish because in nature organisms are unified wholes. There is no history of the giraffe's neck or the elephant's trunk independent of the history of the giraffe or the elephant. Sometimes it is useful to use models that abstract a tiny part of this totality, but we must always remember that these are only models, and that they only tell a partially true part of the truth.

Much of what goes wrong with the application of evolution can be connected to a familiar bugbear of scientific thought, reductionism. Reductionism is the view of science that holds that to understand a thing scientifically we must take it apart and see how the parts fit together, and how the behaviour of the whole derives from the behaviour of the parts. This is a fundamentally important scientific methodology, but it has its limits and these are not always properly respected. Several misuses of reductionism are central to my argument. The first is the attempt to atomize organisms into traits and provide distinct explanations for the evolution of these traits. It can be illuminating to detect a function of a part of an organism that has been important in causing organisms with this part to be selected, but it is essential to be aware that only organisms not their parts are really selected, and most parts of organisms will have many different effects on the success of the organism. The second, and perhaps more classic, abuse of reductionism is the overemphasis on the genetic. Genes are ideal entities for the reductionist, and it is not surprising that extraordinary powers over the organism have been attributed to them, and distorted accounts of evolution have been offered in terms of them. One of the most important steps towards making sense of biology is getting a proper perspective on the extraordinary, but not unlimited, powers of genes.

Both these reductionist errors plague biological attempts to understand human nature. The complexity of the developmental process is greater in the human than in any other case. The human mind develops under continuous and interacting influences from within and without. It is therefore impossible to specify one-to-one relations between elements of the genome and aspects of the mind. The attempt to do so, combined with the reductionistic view of evolution that sees nothing but the accumulation of genes, provides a misguided and worthless route to understanding the human mind. In fact, the hopelessness of the attempt to correlate features of the genome with features of the mind can provide an extreme illustration of the hopelessness of the genetic view of evolution. One cannot abstract a sequence of genomes as the heart of the evolutionary process simply because a genome doesn't specify a pheno type. As an increasing number of theorists are coming to recognize, development must somehow be put back into our view of evolution.

So much for the negative message of the book. On the positive side, there is a great deal of often routine, sometimes exciting, scientific work as the theory is developed and deployed. The study of fossils, the discovery and classification of living forms and their geographic relationships, and the accumulation of knowledge of similarities and differences in the chemistry of life all contribute to a more accurate picture of the actual history of life on our planet. Theoretical work on evolutionary models gives us a better view of what kinds of evolutionary process are possible or likely, and permit informed judgements about the processes that actual lineages may have undergone. There is, in short, a great deal of valuable and largely unproblematic work extending our knowledge of evolution in general and in particular, very much what Thomas Kuhn meant by normal science. I have not said much about this in this book because much of it is of interest mainly to specialists and enthusiasts. I hope I have said enough to indicate the broad areas in which such work is taking place.

I have, on the other hand, emphasized a message from evolution of more general interest, and a message of evolution that I think is often understated. Darwin and his intellectual descendants have provided us with fundamental insight into the nature of the world we live in and of our place within it, a contribution to our basic metaphysics. It is still widely supposed that this is the sort of thing that should come from philosophers or even theologians. In this case, at any rate, the insight has come from biology and I, as a philosopher, am happy just to do my best to interpret it. The theologians, I have suggested, can be less complacent about this insight, and may even need to retrain for a discipline with a subject matter with stronger claims to existence.

In short, I want to insist on a view that once seemed obvious but may now seem naive. This is the view that a great part of Darwin's contribution was to take a major step down the road leading away from primitive animism through the great scientific heroes of the Renaissance—Copernicus, Galileo, Newton, and others—to a naturalistic world-view that was able finally and fully to dispense with the ghosts, spirits, and gods that served to explain natural phenomena for an earlier age.

I said that this picture may seem naive, because in recent decades we have learned to be much more sceptical about the claims of science. And indeed this book itself illustrates some of that scepticism. Certainly, I have been critical of some ideas that pass for scientific and have a great deal of public currency. My point is not to claim that science has told us everything important about the world, that there are no longer any mysteries yet to be discovered, or even that science can ever tell us everything we would like to know. I have no doubt that there are more things in Heaven and Earth than are dreamed of in anyone's philosophy. My point is rather that we know enough to accept our ignorance. We have enough idea of how we can, sometimes, find out even quite profound truths about the world we inhabit that we should no longer be satisfied with mythologies that are made up from sheer ignorance. And that is the real force of my earlier insistence on empiricism. My brand of empiricism does not insist that we must have fully compelling grounds for the things we believe, or indeed that we can find totally irresistible grounds for anything much beyond the immediate and banal. It insists only that we have some reason for the things we believe and that we decline to believe those things for which we have no reasons. A modest requirement, perhaps, but one that would dispose, I contend, with a large part of the religious and superstitious mythologies that continue to dominate and sometimes devastate human lives.

Further Reading

Chapter 1

Karl Popper's most famous work on methodology is *The Logic of Scientific Discovery* (London: Hutchinson, 1959). The discussion of shrikes and the use to which it is put are drawn from David Buss, *The Evolution of Desire* (New York: Basic Books, 1994). Also referred to is Ben Greenstein, *The Fragile Male* (New York: Birch Lane Press, 1993).

Chapter 2

There are several good general introductions to the theory of evolution. John Maynard Smith, *The Theory of Evolution* (Harmondsworth: Penguin, 1977), is among the best. A more philosophical treatment, which goes into considerable depth on the topic of natural selection, is Elliott Sober, *The Nature of Selection* (Cambridge, Mass.: MIT Press, 1984). Sober's book develops a pluralistic view of the units of selection. More recently he has been involved in the attempt to rehabilitate the importance of group selection, Elliott Sober and David Sloan Wilson, *Unto Others: The Evolution of Altruism* (Cambridge, Mass.: Harvard University Press, 1998). Richard Dawkins's classic exposition of gene selectionism is *The Selfish Gene* (Oxford: Oxford University Press, 1976). He has subsequently written a series of extremely engaging and readable books on evolution.

The classic source for developmental systems theory is Susan

Oyama, *The Ontogeny of Information* (Cambridge: Cambridge University Press, 1985). This is not an easy book, however. A lucid overview can be found in P. E. Griffiths and R. D. Gray, 'Developmental Systems and Evolutionary Explanations', *Journal of Philosophy*, 91 (1994), 277–304. A recent and excellent book that develops ideas from this tradition is Lenny Moss, *What Genes Can't Do* (Cambridge, Mass.: MIT Press, 2002).

Punctuated equilibrium theory was introduced by Niles Eldredge and Stephen Jay Gould in 'Punctuated Equilibria: An Alternative to Phyletic Gradualism', in T. Schopf (ed.), *Models in Paleobiology* (San Francisco: Freeman, 1972).

Chapter 3

A standard source for the classic model of explanation by derivation is Carl G. Hempel, *Philosophy of Natural Science* (Englewood Cliffs, NJ: Prentice-Hall, 1968/1999). David Hume's philosophy is approached most easily through his *An Enquiry Concerning Human Understanding* (1748).

A great deal of sceptical thought about the possibility of providing detailed adaptive explanations derives from Richard Lewontin, *The Genetic Basis of Evolutionary Change* (New York: Columbia University Press, 1974). A classic article on the topic is S. J. Gould and R. C. Lewontin, 'The Spandrels of San Marco and the Panglossian Paradigm: A Critique of the Adaptationist Programme', *Proceedings of the Royal Society of London*, 205 (1979), 581–98.

The distinction between adaptation and exaptation was introduced by S. J. Gould and Elizabeth Vrba, 'Exaptation—a Missing Term in the Science of Form', *Paleobiology*, 8 (1982), 4–15. Gould on the panda's thumb can be found in *The Panda's Thumb* (New York: Norton, 1980). Other delightful essays illustrating the general theme can be found in this volume and other volumes of Gould's essays.

The classic source for the use of models in science is Mary Hesse, *Models and Analogies in Science* (South Bend, Ind.: University of Notre Dame Press, 1966). For a recent development of the view of science as involving the construction of models rather than the discovery of laws, see Ronald Giere, *Science Without Laws* (Chicago: University of Chicago Press, 1999).

Chapter 4

There have been a number of excellent books published on Darwin in recent years. The one that most specifically addresses his relation to religion is James Moore, *The Darwin Legend* (London: Hodder and Stoughton, 1995). For Darwin's account of himself, see *The Autobiography of Charles Darwin*, ed. N. Barlow (London: Collins, 1958).

William James's pragmatic account of belief is presented in his essay 'The Will to Believe', in a collection of essays of the same name (Cambridge, Mass.: Harvard University Press, 1979). The classic statement of the argument from design is William Paley, *Natural Theology* (1802). Hume's critique of this argument is in *Dialogues Concerning Natural Religion* (1779).

An example of the new version of the design argument that goes under the banner of intelligent design is William A. Dembski, *Intelligent Design: The Bridge Between Science and Theology* (Downers Grove, Ill.: InterVarsity Press, 1999). Inference to the best explanation is discussed by Peter Lipton,

Inference to the Best Explanation (London: Routledge, 2003). The works discussed in the text reconciling evolution and Christianity are Michael Ruse, *Can a Darwinian be a Christian?* (Cambridge: Cambridge University Press, 2001), and Gould, *Rocks of Ages* (New York: Vintage Books, 2002).

Stuart Kauffman's views on self-organization are most accessible in *At Home in the Universe* (Oxford: Oxford University Press, 1995). A good discussion of current ideas on the origin of life can be found in John Maynard Smith and Eors Szathmary, *The Origins of Life: From the Birth of Life to the Origins of Language* (Oxford: Oxford University Press, 2000).

Chapter 5

René Descartes' dualism is described in *Meditations on First Philosophy* (1642). A classic and energetic rebuttal is Gilbert Ryle, *The Concept of Mind* (London: Hutchinson, 1949). A rare contemporary defender of the view that non-human animals lack consciousness is Peter Carruthers, 'Brute Experience', *Journal of Philosophy*, 86 (1989), 258–69. My own opposing view can be found in *Humans and Other Animals* (Oxford: Oxford University Press, 2002), Ch. 10. Difficulties with the general conception of life after death are discussed by Bernard Williams, *Problems of the Self* (Cambridge: Cambridge University Press, 1973), Chs. 1–5. Ch. 6 explains why it would not be much fun.

Communication in Vervet monkeys is described by Dorothy Cheney and Robert Seyfarth, *How Monkeys See the World* (Chicago: Chicago University Press, 1990). Experiments on teaching language to great apes are discussed in my *Humans and Other Animals*, Ch. 11, which includes citations to some of the most important original research. A readable account of the evolution of language is Steven Pinker, *The Language Instinct* (Harmondsworth: Penguin, 1995).

Chapter 6

Thomas Kuhn's classic work is *The Structure of Scientific Revolutions* (Chicago: University of Chicago Press, 1962). E. O. Wilson's opus magnus was *Sociobiology: The New Synthesis* (Cambridge, Mass.: Harvard University Press, 1975). Sustained critiques were provided in Richard Lewontin, Steven Rose, and Leon Kamin, *Not in our Genes: Biology, Ideology, and Human Nature* (New York: Pantheon, 1984); and Philip Kitcher, *Vaulting Ambition: Sociobiology and the Quest for Human Nature* (Cambridge, Mass.: MIT Press, 1985). The history of the controversy over sociobiology is well related by Ullica Segerstrale, *Defenders of the Truth* (Oxford: Oxford University Press, 2000). Key documents, including the initial attacks on Wilson by the Sociology Study Group of Science for the People and Wilson's replies, can be found in Arthur Caplan (ed.), *The Sociobiology Debate* (New York: Harper and Row, 1978). A definitive statement of the aims of evolutionary psychology is Jerome Barkow, Leda Cosmides, and John Tooby (eds.), *The Adapted Mind* (Oxford: Oxford University Press, 1992), especially the very long introductory essay by Tooby and Cosmides. A popular treatment is Steven Pinker, *How the Mind Works* (New York: Norton, 1997). Extensive criticism can be found in Hilary Rose and Steven Rose (eds.), *Alas, Poor Darwin* (London: Jonathan Cape, 2000) and my *Human Nature and the Limits of Science* (Oxford: Oxford University Press, 2001).

The reference in the text to Dennett is to Daniel Dennett's

Darwin's Dangerous Idea (New York: Simon and Schuster, 1995). A similar appropriation of theological thought can be seen in the reference to Paley in Richard Dawkins's title, *The Blind Watchmaker* (New York: Norton, 1986). The status of and evidence for human behavioural genetics is very well explained by Jonathan Kaplan, *The Limits and Lies of Human Genetic Research* (London: Routledge, 2000) which also provides forceful criticism of many unwarranted claims that derive from this source.

Interest in evolutionary psychology's speculations on rape have been stirred up by Randy Thornhill and Craig Palmer, *A Natural History of Rape: Biological Bases of Sexual Coercion* (Cambridge, Mass.: MIT Press, 2000). A range of excellent rebuttals can be found in Cheryl Brown Travis (ed.), *Evolution, Gender, and Rape* (Cambridge, Mass.: MIT Press, 2003).

Chapter 7

The genetic basis for racial difference, or lack of it, is discussed in Stephen Jay Gould, *The Mismeasure of Man* (New York: Norton, 1996) and Lewontin, Rose, and Kamin, *Not in our Genes*. The essay by Pigliucci and Kaplan can be found at
http://fp.bio.utk.edu/wisdom/Essays/race.html.

The most notorious recent source for claims about the correlation of race and IQ is Richard Herrnstein and Charles Murray, *The Bell Curve: Intelligence and Class Structure in American Life* (New York: Simon and Schuster, 1994). A powerful critique of their argument . . . ed Block, 'How Heritability Misleads about Race', *Cognition*, 56 (1995), 90–128. See also discussion in Kaplan, *The Limits and Lies of Human Genetic Research*. Anne Fausto-Sterling, *Sexing the Body: Gender Politics and the Construction of Sexuality*

(New York: Basic Books, 2000) gives an illuminating discussion of the enforcement of sexual dimorphism. The economic analysis of sex roles in terms of parental investment is generally attributed to Robert Trivers, 'Parental Investment and Sexual Selection', in B. Campbell (ed.), *Sexual Selection and the Descent of Man* (London: Heinemann, 1972). The person who has most energetically attempted to investigate the gender differences predicted by evolutionary psychology is David Buss, *The Evolution of Desire* (New York: Basic Books, 1994). The project is criticized in greater detail in my *Human Nature and the Limits of Science* (Oxford: Oxford University Press, 2001).

A summary of bonobo culture by the eminent primatologist Frans de Waal can be found at **http://songweaver.com/info/bonobos.html.** (This was originally published in *Scientific American* (March 1995), 82–8.)

Chapter 8

Reductionism is criticized in detail in my *The Disorder of Things: Metaphysical Foundations of the Disunity of Science* (Cambridge, Mass.: Harvard University Press, 1993).

Index